The Gallic War & the Civil War

Julius Caesar

translated by W. A. MacDevitt

TITLE
Commentarii de Bello Gallico
& Commentarii de Bello Civili

AUTHOR
Julius Caesar

YEARS PUBLISHED
58-49 & 46 BCE

LANGUAGE
Latin

* * *

TITLE
"De Bello Gallico" & Other Commentaries

TRANSLATOR
W. A. MacDevitt

YEAR PUBLISHED
1915

LANGUAGE
English

The Gallic War & the Civil War

Foundations of Fantasy Roleplaying Games

A series exploring the literature that influenced the modern genres of fantasy, science fiction, and horror, and the roleplaying games they continue to inspire.

This series is dedicated to all the hardworking game masters the world round. May you and your players find in these books more inspiration for your games.

While this series orients itself towards genre fiction and roleplaying games, they are also for general readers desiring quality copies of public domain works.

Published by
Charybdis Press
charybdispress.com

CP0013
2024

Some editorial changes were made to the original text.
To help improve the copy, email info@charybdispress.com
subject line: "title" text, if any typos or other errata are found.

This is a work from the past written with different
sensitivities and sensibilities to the modern reader.
It should go without saying, but CAVEAT LECTOR.

Editing, Layout, Design: J. Blasso-Gieseke

ISBN 978-1-957399-28-7

FRONT COVER ART
Adolphe Yvon, 1875, oil on canvas, *César franchit le Rubicon*
Musée des Beaux-Arts d'Arras (Pas-de-Calais, France), public domain
INTERIOR MAP
Feitscherg, CC Attribution-Share Alike 3.0 Unported

Editor's Introduction

Game Masters: The history of fantasy roleplaying begins with wargaming. And what better chronicle of ancient war exists than one penned by the hand of Caesar himself? What's of interest here are the tactics, earthworks, siege equipment, reconnaisance, movement, positioning, alliances, troop type, supplies, season, weather, terrain, morale, which provides a general overview of war in all of its dynamic complexity. If you are unfamiliar with any of these, this an excellent introduction. If you are already familiar with them, this is an excellent review. Reading this, it is easy to imagine an invading or defending monster army. Some roleplaying games offer systems for controlling armies, becoming near analogues of wargames. However, most do not, and since roleplaying games are designed to be played with a small party of PCs, this is often the best scale to keep the game at. Where this book may help is if you're desigining a scenario where PCs must get behind enemy lines to assassinate the leaders of an invading army or to gain and hold a position of critical tactical advantage or are in a besieged castle, town, or village. Whatever the scenario, introducing large scale warfare or small, deadly skirmishes may be a way to shake things up from the regular dungeon delves and that, if they win, include the PCs in a part of local history and lore.

Author's Introduction

The character of the First Caesar has perhaps never been worse appreciated than by him who in one sense described it best; that is, with most force and eloquence wherever he really *did* comprehend it. This was Lucan, who has nowhere exhibited more brilliant rhetoric, nor wandered more from the truth, than in the contrasted portraits of Caesar and Pompey. The famous line, "*Nil actum reputans si quid superesset agendum,*" is a fine feature of the real character, finely expressed. But, if it had been Lucan's purpose (as possibly, with a view to Pompey's benefit, in some respects it was) utterly and extravagantly to falsify the character of the great Dictator, by no single trait could he more effectually have fulfilled that purpose, nor in fewer words, than by this expressive passage, "*Gaudensque viam fecisse ruina.*" Such a trait would be almost extravagant applied even to Marius, who (though in many respects a perfect model of Roman grandeur, massy, columnar, imperturbable, and more perhaps than any one man recorded in History capable of justifying the bold illustration of that character in Horace, "*Si fractus illabatur orbis, impavidum ferient ruinae*") had, however, a ferocity in his character, and a touch of the devil in him, very rarely united with the same tranquil intrepidity. But, for Caesar, the all-accomplished statesman, the splendid orator, the man of elegant habits and polished taste, the patron of the fine arts in a degree transcending all example of his own or the previous age, and as a man of general literature so much beyond his contemporaries, except Cicero, that he looked down even upon the brilliant Sylla as an illiterate person — to class such a man with the race of furious destroyers exulting in the desolations they spread is to err not by an individual trait, but by the whole genus. The Attilas and the Tamerlanes, who rejoice in avowing themselves the scourges of God, and the special instruments of his wrath, have no one feature of affinity to the polished and humane Caesar, and would as little have comprehended his character as he could have respected theirs. Even Cato, the unworthy hero of Lucan, might have suggested to him a little more truth in this instance, by a celebrated remark which he made on the characteristic distinction of Caesar, in comparison with other revolutionary disturbers; for, said he, whereas others had attempted the overthrow of the state in a continued paroxysm of fury, and in a state of mind resembling the lunacy of intoxication, Caesar, on the contrary, among that whole class of civil disturbers, was the only one who had come to the task in a temper of sobriety and moderation (*unum accessisse sobrium ad rempublicam delendam*)....

Great as Caesar was by the benefit of his original nature, there can be no doubt that he, like others, owed something to circumstances; and perhaps amongst those which were most favourable to the premature development of great self-dependence we must reckon the early death of his father. It is, or it is not, according to the nature of men, an advantage to be orphaned at as early age. Perhaps utter orphanage is rarely or never such: but to lose a father betimes may, under appropriate circumstances, profit a strong mind greatly. To Caesar it was a prodigious benefit that he lost his father when not much more than fifteen. Perhaps it was an advantage also to his father that he died thus early. Had he stayed a year longer, he might have seen himself despised, baffled, and made ridiculous. For where, let us ask, in any age, was the father capable of adequately sustaining that relation to the unique Caius Julius — to him, in the appropriate language of Shakespeare

"The foremost man of all this world?"

And, in this fine and Caesarean line, "this world" is to be understood not of the order of co-existences merely, but also of the order of successions; he was the foremost man not only of his contemporaries, but also, within his own intellectual class, of men generally — of all that ever should come after him, or should sit on thrones under the denominations of Czars, Kesars, or Caesars of the Bosphorus and the Danube; of all in every age that should inherit his supremacy of mind, or should subject to themselves the generations of ordinary men by qualities analogous to his. Of this infinite superiority some part must be ascribed to his early emancipation from paternal control. There are very many cases in which, simply from considerations of sex, a female cannot stand forward as the head of a family, or as its suitable representative. If they are even ladies paramount, and in situations of command, they are also women. The staff of authority does not annihilate their sex; and scruples of female delicacy interfere for ever to unnerve and emasculate in their hands the sceptre however otherwise potent. Hence we see, in noble families, the merest boys put forward to represent the family dignity, as fitter supporters of that burden than their mature mothers. And of Caesar's mother, though little is recorded, and that little incidentally, this much at least we learn — that, if she looked down upon him with maternal pride and delight, she looked up to him with female ambition as the re-edifier of her husband's honours, — looked with reverence as to a column of the Roman grandeur and with fear and feminine anxieties as to one whose aspiring spirit carried him but too prematurely into the fields of adventurous strife. One slight and evanescent sketch of the relations which subsisted between Caesar and his mother, caught from the wrecks of time, is preserved both by Plutarch and Suetonius. We see in the early dawn the young patrician standing upon the steps of his patrimonial portico, his mother with her arms wreathed about his neck, looking up to his noble countenance, sometimes drawing auguries of hope from features so fitted for command, sometimes boding an early blight to promises so dangerously magnificent. That she had something of her son's aspiring character, or that he presumed so much in a mother of his, we learn from the few words which survive of their conversation. He addressed to her no language that could tranquillise her fears. On the contrary, to any but a Roman mother his valedictory words, taken in connexion with the known determination of his character, were of a nature to consummate her depression, as they tended to confirm the very worst of her fears. He was then going to stand his chance in a popular electioneering contest for an office of the highest dignity, and to launch himself upon the storms of the Campus Martius. At that period, besides other and more ordinary dangers, the bands of gladiators, kept in the pay of the more ambitious or turbulent amongst the Roman nobles, gave a popular tone of ferocity and of personal risk to the course of such contests; and, either to forestall the victory of an antagonist, or to avenge their own defeat, it was not at all impossible that a body of incensed competitors might intercept his final triumph by assassination. For this danger, however, he had no leisure in his thoughts of consolation; the sole danger which he contemplated, or supposed his mother to contemplate, was the danger of defeat, and for that he reserved his consolations. He bade her fear nothing; for that his determination was to return with victory, and with the ensigns of the dignity he sought, or to return a corpse.

Early indeed did Caesar's trials commence; and it is probable, that, had not the death of his father, by throwing him prematurely upon his own resources, prematurely developed the masculine features of his character, forcing him whilst yet a boy under the discipline of civil conflict and the yoke of practical life, even his energies might have been insufficient to sustain them. His age is not exactly ascertained; but it is past a doubt that he had not reached his twentieth year when he had the hardihood to engage in a struggle with Sylla, then Dictator, and exercising the immoderate powers of that office with the licence and the severity which History has made so memorable. He had neither any distinct grounds of hope, nor any eminent example at that time, to countenance him in this struggle — which yet he pushed on in the most uncompromising style, and to the utmost verge of defiance. The subject of the contest gives it a further interest. It was the youthful wife of the youthful Caesar who stood under the shadow of the great Dictator's displeasure; not personally, but politically, on account of her connexions: and her it was, Cornelia, the daughter of a man who had been four times consul, that Caesar was required to divorce: but he spurned the haughty mandate, and carried his determination to a triumphant issue, notwithstanding his life was at stake, and at one time saved only by shifting his place of concealment every night; and this young lady it was who afterwards became the mother of his only daughter. Both mother and daughter, it is remarkable, perished prematurely, and at critical periods of Caesar's life; for it is probable enough that these irreparable wounds to Caesar's domestic affections threw him with more exclusiveness of devotion upon the fascinations of glory and ambition than might have happened under a happier condition of his private life. That Caesar should have escaped destruction in this unequal contest with an enemy then wielding the whole thunders of the state, is somewhat surprising; and historians have sought their solution of the mystery in the powerful intercessions of the vestal virgins, and several others of high rank amongst the connexions of his great house. These may have done something; but it is due to Sylla, who had a sympathy with everything truly noble, to suppose him struck with powerful admiration for the audacity of the young patrician, standing out in such severe solitude among so many examples of timid concession; and that to this magnanimous feeling in the Dictator much of the indulgence which he showed may have been really due. In fact, according to some accounts, it was not Sylla, but the creatures of Sylla (*adjutores*), who pursued Caesar. We know, at all events, that Sylla formed a right estimate of Caesar's character, and that, from the complexion of his conduct in this one instance, he drew that famous prophecy of his future destiny; bidding his friends beware of that slipshod boy, "for that in him lay couchant many a Marius." A grander testimony to the awe which Caesar inspired, or from one who knew better the qualities of that Cyclopean man by whose scale he measured the patrician boy, cannot be imagined.

It is not our intention, or consistent with our plan, to pursue this great man through the whole circumstances of his romantic career; though it is certain that many parts of his life require investigation much keener than has ever been applied to them, and that many might be placed in a new light. Indeed, the whole of this most momentous section of ancient history ought to be recomposed with the critical scepticism of a Niebuhr, and the same comprehensive collation, resting, if possible, on the felicitous interpretation of authorities. In reality it is the hinge upon which turned the future destiny of the whole earth, and, having therefore a common relation to all modern nations whatsoever, should naturally have been cultivated with the zeal which belongs to a personal concern. In general, the anecdotes which express most vividly the grandeur of character in the first Caesar are those which illustrate his defiance of danger in extremity: the prodigious energy and rapidity of his decisions and motions in the field (looking to which it was that Cicero called him τέρας [*teras* Greek: monster] or portentous revelation); the skill with which he penetrated the designs of his enemies, and the electric speed with which he met disasters with remedy and reparation, or, where that was impossible, with relief; the extraordinary presence of mind which he showed in turning adverse omens to his own advantage, as when, upon stumbling in coming on shore (which was esteemed a capital omen of evil), he transfigured as it were in one instant its whole meaning by exclaiming, "Thus, and by this contact with the earth, do I take possession of thee, O Africa!" in that way giving to an accident the semblance of a symbolic purpose. Equally conspicuous was the grandeur of fortitude with which he faced the whole extent of a calamity when palliation could do no good, *"non negando, minuendove, sed insuper amplificando, ementiendoque"*; as when, upon finding his soldiery alarmed at the approach of Juba, with forces really great, but exaggerated by their terrors, he addressed them in a military harangue to the following effect: — "Know that within a few days the king will come up with us, bringing with him sixty thousand legionaries, thirty thousand cavalry, one hundred thousand light troops, besides three hundred elephants. Such being the case, let me hear no more of conjectures and opinions, for you have now my warrant for the fact, whose information is past doubting. Therefore, be satisfied; otherwise, I will put every man of you on board some crazy old fleet, and whistle you down the tide — no matter under what winds, no matter towards what shore." Finally, we might seek for *characteristic* anecdotes of Caesar in his unexampled liberalities and contempt of money.

Upon this last topic it is the just remark of Casaubon that some instances of Caesar's munificence have been thought apocryphal, or to rest upon false readings, simply from ignorance of the heroic scale upon which the Roman splendours of that age proceeded. A forum which Caesar built out of the products of his last campaign, by way of a present to the Roman people, cost him — for the ground merely on which it stood — nearly eight hundred thousand pounds. To the citizens of Rome he presented, in one *congiary*, about two guineas and a half a head. To his army, in one *donation*, upon the termination of the Civil War, he gave a sum which allowed about two hundred pounds a man to the infantry, and four hundred to the cavalry. It is true that the legionary troops were then much reduced by the sword of the enemy, and by the tremendous hardships of their last campaigns. In this, however, he did perhaps no more than repay a debt. For it is an instance of military attachment, beyond all that Wallenstein or any commander, the most beloved amongst his troops, has ever experienced, that, on the breaking out of the Civil War, not only did the centurions of every legion severally maintain a horse soldier, but even the privates volunteered to serve without pay, and (what might seem impossible) without their daily rations. This was accomplished by subscriptions amongst themselves, the more opulent undertaking for the maintenance of the needy. Their disinterested love for Caesar appeared in another and more difficult illustration: it was a traditionary anecdote in Rome that the majority of those amongst Caesar's troops who had the misfortune to fall into the enemy's hands refused to accept their lives under the condition of serving against *him*.

In connexion with this subject of his extraordinary munificence, there is one aspect of Caesar's life which has suffered much from the misrepresent-ations of historians, and that is — the vast pecuniary embarrassments under which he laboured, until the profits of war had turned the scale even more prodigiously in his favour. At one time of his life, when appointed to a foreign office, so numerous and so clamorous were his creditors that he could not have left Rome on his public duties had not Crassus come forward with assistance in money, or by guarantees, to the amount of nearly two hundred thousand pounds. And at another he was accustomed to amuse himself with computing how much money it would require to make him worth exactly nothing (*i.e.* simply to clear him of debts); this, by one account, amounted to upwards of two millions sterling. Now, the error of historians has been to represent these debts as the original ground of his ambition and his revolutionary projects, as though the desperate condition of his private affairs had suggested a civil war to his calculations as the best or only mode of redressing it. Such a policy would have resembled the last desperate resource of an unprincipled gambler, who, on seeing his final game at chess, and the accumulated stakes depending upon it, all on the brink of irretrievable sacrifice, dexterously upsets the chess-board, or extinguishes the lights. But Julius, the one sole patriot of Rome, could find no advantage to his plans in darkness or in confusion. Honestly supported, he would have crushed the oligarchies of Rome by crushing in its lairs that venal and hunger-bitten democracy which made oligarchy and its machineries resistless. Caesar's debts, far from being stimulants and exciting causes of his political ambition, stood in an inverse relation to the ambition; they were its results, and represented its natural costs, being contracted from first to last in the service of his political intrigues, for raising and maintaining a powerful body of partisans, both in Rome and elsewhere. Whosoever indeed will take the trouble to investigate the progress of Caesar's ambition, from such materials as even yet remain, may satisfy himself that the scheme of revolutionizing the Republic, and placing himself at its head, was no growth of accident or circumstances; above all, that it did not arise upon any so petty and indirect a suggestion as that of his debts; but that his debts were in their very first origin purely ministerial to his wise, indispensable, and patriotic ambition; and that his revolutionary plans were at all periods of his life a direct and foremost object, but in no case bottomed upon casual impulses. In this there was not only patriotism, but in fact the one sole mode of patriotism which could have prospered, or could have found a field of action.

Chatter not, sublime reader, commonplaces of scoundrel moralists against ambition. In some cases ambition is a hopeful virtue; in others (as in the Rome of our resplendent Julius) ambition was the virtue by which any other could flourish. It had become evident to everybody that Rome, under its present constitution, must fall; and the sole question was — by whom? Even Pompey, not by nature of an aspiring turn, and prompted to his ambitious course undoubtedly by circumstances and, the friends who besieged him, was in the habit of saying, *"Sylla potuit: ego non potero?"* Sylla found it possible: shall I find it not so? Possible to do what? To overthrow the political system of the Republic. This had silently collapsed into an order of things so vicious, growing also so hopelessly worse, that all honest patriots invoked a purifying revolution, even though bought at the heavy price of a tyranny, rather than face the chaos of murderous distractions to which the tide of feuds and frenzies was violently tending.

Such a revolution at such a price was not less Pompey's object than Caesar's. In a case, therefore, where no benefit of choice was allowed to Rome as respected the thing, but only as respected the person, Caesar had the same right to enter the arena in the character of combatant as could belong to any one of his rivals. And that he *did* enter that arena constructively, and by secret design, from his very earliest manhood, may be gathered from this — that he suffered no openings towards a revolution, provided they had any hope in them, to escape his participation. It is familiarly known that he was engaged pretty deeply in the conspiracy of Catiline, and that he incurred considerable risk on that occasion; but it is less known that he was a party to at least two other conspiracies. There was even a fourth, meditated by Crassus, which Caesar so far encouraged as to undertake a journey to Rome from a very distant quarter merely with a view to such chances as it might offer to him; but, as it did not, upon examination, seem to him a very promising scheme, he judged it best to look coldly upon it, or not to embark in it by any personal co-operation. Upon these and other facts we build our inference — that the scheme of a revolution was the one great purpose of Caesar from his first entrance upon public life. Nor does it appear that he cared much by whom it was undertaken, provided only there seemed to be any sufficient resources for carrying it through, and for sustaining the first collision with the regular forces of the existing oligarchies, taking or not taking the shape of triumvirates. He relied, it seems, on his own personal superiority for raising him to the head of affairs eventually, let who would take the nominal lead at first.

To the same result, it will be found, tended the vast stream of Caesar's liberalities. From the senator downwards to the lowest *faex Romuli*, he had a hired body of dependents, both in and out of Rome, equal in numbers to a nation. In the provinces, and in distant kingdoms, he pursued the same schemes. Everywhere he had a body of mercenary partisans; kings even are known to have taken his pay. And it is remarkable that even in his character of commander-in-chief, where the number of legions allowed to him for the accomplishment of his Gaulish mission raised him for a number of years above all fear of coercion or control, he persevered steadily in the same plan of providing for the distant day when he might need assistance, not *from* the state, but *against* the state. For, amongst the private anecdotes which came to light under the researches made into his history after his death, was this — that, soon after his first entrance upon his government in Gaul, he had raised, equipped, disciplined, and maintained, from his own private funds, a legion amounting, possibly, to six or seven thousand men, who were bound to no sacrament of military obedience to the state, nor owed fealty to any auspices except those of Caesar. This legion, from the fashion of their crested helmets, which resembled the heads of a small aspiring bird, received the popular name of the *Alauda* (or Lark) legion. And very singular it was that Cato, or Marcellus, or some amongst those enemies of Caesar who watched his conduct during the period of his Gaulish command with the vigilance of rancorous malice, should not have come to the knowledge of this fact; in which case we may be sure that it would have been denounced to the Senate.

Such, then, for its purpose and its uniform motive, was the sagacious munificence of Caesar. Apart from this motive, and considered in and for itself, and simply with a reference to the splendid forms which it often assumed, this munificence would furnish the materials for a volume. The public entertainments of Caesar, his spectacles and shows, his *naumachiae*, and the pomps of his unrivalled triumphs (the closing triumphs of the Republic), were severally the finest of their kind which had then been brought forward. Sea-fights were exhibited upon the grandest scale, according to every known variety of nautical equipment and mode of conflict, upon a vast lake formed artificially for that express purpose. Mimic land-fights were conducted, in which all the circumstances of real war were so faithfully rehearsed that even elephants "indorsed with towers," twenty on each side, took part in the combat. Dramas were represented in every known language (*per omnium linguarum histriones*). And hence (that is, from the conciliatory feeling thus expressed towards the various tribes of foreigners resident in Rome) some have derived an explanation of what is else a mysterious circumstance amongst the ceremonial observances at Caesar's funeral — that all people of foreign nations then residing at Rome distinguished themselves by the conspicuous share which they took in the public mourning; and that, beyond all other foreigners, the Jews for night after night kept watch and ward about the Emperor's grave. Never before, according to traditions which lasted through several generations in Rome, had there been so vast a conflux of the human race congregated to any one centre, on any one attraction of business or of pleasure, as to Rome on occasion of these triumphal spectacles exhibited by Caesar.

In our days, the greatest occasional gatherings of the human race are in India, especially at the great fair of the *Hurdwar* on the Ganges in northern Hindustan: a confluence of some millions is sometimes seen at that spot, brought together under the mixed influences of devotion and commercial business, but very soon dispersed as rapidly as they had been convoked. Some such spectacle of nations crowding upon nations, and some such Babylonian confusion of dresses, complexions, languages, and jargons, was then witnessed at Rome. Accommodations within doors, and under roofs of houses, or roofs of temples, was altogether impossible. Myriads encamped along the streets, and along the high-roads, fields, or gardens. Myriads lay stretched on the ground, without even the slight protection of tents, in a vast circuit about the city. Multitudes of men, even senators, and others of the highest rank, were trampled to death in the crowds. And the whole family of man might seem at that time to be converged at the bidding of the dead Dictator. But these, or any other themes connected with the public life of Caesar, we notice only in those circumstances which have been overlooked, or partially represented, by historians. Let us now, in conclusion, bring forward, from the obscurity in which they have hitherto lurked, the anecdotes which describe the habits of his private life, his tastes, and personal peculiarities.

In person, he was tall, fair, gracile, and of limbs distinguished for their elegant proportions. His eyes were black and piercing. These circumstances continued to be long remembered, and no doubt were constantly recalled to the eyes of all persons in the imperial palaces by pictures, busts, and statues; for we find the same description of his personal appearance three centuries afterwards in a work of the Emperor Julian's. He was a most accomplished horseman, and a master (*peritissimus*) in the use of arms. But, notwithstanding his skill and horsemanship, it seems that, when he accompanied his army on marches, he walked oftener than he rode; no doubt, with a view to the benefit of his example, and to express that sympathy with his soldiers which gained him their hearts so entirely. On other occasions, when travelling apart from his army, he seems more frequently to have ridden in a carriage than on horseback. His purpose, in this preference, must have been with a view to the transport of luggage. The carriage which he generally used was a *rheda*, a sort of gig, or rather curricle; for it was a *four*-wheeled carriage, and adapted (as we find from the imperial regulations for the public carriages, etc.) to the conveyance of about half a ton. The mere personal baggage which Caesar carried with him was probably considerable; for he was a man of elegant habits, and in all parts of his life sedulously attentive to elegance of personal appearance. The length of journeys which he accomplished within a given time appears even to us at this day, and might well therefore appear to his contemporaries, truly astonishing. A distance of one hundred miles was no extraordinary day's journey for him in a *rheda*, such as we have described it. So refined were his habits, and so constant his demand for the luxurious accommodations of polished life as it then existed in Rome, that he is said to have carried with him, as indispensable parts of his personal baggage, the little ivory lozenges, squares and circles or ovals, with other costly materials, wanted for the tessellated flooring of his tent. Habits such as these will easily account for his travelling in a carriage rather than on horseback.

The courtesy and obliging disposition of Caesar were notorious; and both were illustrated in some anecdotes which survived for generations in Rome. Dining on one occasion, as an invited guest, at a table where the servants had inadvertently, for salad-oil, furnished some sort of coarse lamp-oil, Caesar would not allow the rest of the company to point out the mistake to their host, for fear of shocking him too much by exposing what might have been construed into inhospitality. At another time, whilst halting at a little *cabaret*, when one of his retinue was suddenly taken ill, Caesar resigned to his use the sole bed which the house afforded. Incidents as trifling as these express the urbanity of Caesar's nature; and hence one is the more surprised to find the alienation of the Senate charged, in no trifling degree, upon a gross and most culpable failure in point of courtesy. Caesar, it is alleged — but might we presume to call upon antiquity for its authority? — neglected to rise from his seat, on their approaching him with an address of congratulation. It is said, and we can believe it, that he gave deeper offence by this one defect in a matter of ceremonial observance than by all his substantial attacks upon their privileges. What we find it difficult to believe is not that result from that offence — this is no more than we should all anticipate — not *that*, but the possibility of the offence itself, from one so little arrogant as Caesar, and so entirely a man of the world. He was told of the disgust which he had given; and we are bound to believe his apology, in which he charged it upon sickness, that would not at the moment allow him to maintain a standing attitude. Certainly the whole tenor of his life was not courteous only, but kind, and to his enemies merciful in a degree which implied so much more magnanimity than men in general could understand that by many it was put down to the account of weakness.

Weakness, however, there was none in Caius Caesar; and, that there might be none, it was fortunate that conspiracy should have cut him off in the full vigour of his faculties, in the very meridian of his glory, and on the brink of completing a series of gigantic achievements. Amongst these are numbered: — a digest of the entire body of laws, even then become unwieldy and oppressive; the establishment of vast and comprehensive public libraries, Greek as well as Latin; the chastisement of Dacia (that needed a cow-hiding for insolence as much as Afghanistan from us in

1840); the conquest of Parthia; and the cutting a ship canal through the Isthmus of Corinth. The reformation of the Calendar he had already accomplished. And of all his projects it may be said that they were equally patriotic in their purpose and colossal in their proportions.

As an orator, Caesar's merit was so eminent that, according to the general belief, had he found time to cultivate this department of civil exertion, the received supremacy of Cicero would have been made questionable, or the honour would have been divided. Cicero himself was of that opinion, and on different occasions applied the epithet *splendidus* to Caesar, as though in some exclusive sense, or with some peculiar emphasis, due to him. His taste was much simpler, chaster, and less inclined to the *florid* and Asiatic, than that of Cicero. So far he would, in that condition of the Roman culture and feeling, have been less acceptable to the public; but, on the other hand, he would have compensated this disadvantage by much more of natural and Demosthenic fervour.

In literature, the merits of Caesar are familiar to most readers. Under the modest title of *Commentaries*, he meant to offer the records of his Gallic and British campaigns, simply as notes, or memoranda, afterwards to be worked up by regular historians; but, as Cicero observes, their merit was such in the eyes of the discerning that all judicious writers shrank from the attempt to alter them. In another instance of his literary labours he showed a very just sense of true dignity. Rightly conceiving that everything patriotic was dignified, and that to illustrate or polish his native language was a service of real and paramount patriotism, he composed a work on the grammar and orthoepy of the Latin language. Cicero and himself were the only Romans of distinction in that age who applied themselves with true patriotism to the task of purifying and ennobling their mother tongue. Both were aware of a transcendent value in the Grecian literature as it then stood; but that splendour did not depress their hopes of raising their own to something of the same level. As respected the natural wealth of the two languages, it was the private opinion of Cicero that the Latin had the advantage; and, if Caesar did not accompany him to that length — which, perhaps, under some limitations he ought to have done — he yet felt that it was but the more necessary to draw forth any special or exceptional advantage which it really had.

Was Caesar, upon the whole, the greatest of men? We restrict the question, of course, to the classes of men great in *action*: great by the extent of their influence over their social contemporaries; great by throwing open avenues to extended powers that previously had been closed; great by making obstacles once vast to become trivial, or prizes that once were trivial to be glorified by expansion. I (said Augustus Caesar) found Rome built of brick; but I left it built of marble. Well, my man, we reply, for a wondrously little chap, you did what in Westmoreland they call a good *darroch* (day's work); and, if navvies had been wanted in those days, you should have had our vote to a certainty. But Caius Julius, even under such a limitation of the comparison, did a thing as much transcending this as it was greater to project Rome across the Alps and the Pyrenees, — expanding the grand Republic into crowning provinces of 1. France (*Gallia*), 2. Belgium, 3. Holland (*Batavia*), 4. England (*Britannia*), 5. Savoy (*Allobroges*), 6. Switzerland (*Helvetia*), 7. Spain (*Hispania*), — than to decorate a street or to found an amphitheatre. Dr. Beattie once observed that, if that question as to the greatest man in action upon the rolls of History were left to be collected from the suffrages already expressed in books and scattered throughout the literature of all nations, the scale would be found to have turned prodigiously in Caesar's favour as against any single competitor; and there is no doubt whatsoever that even amongst his own countrymen, and his own contemporaries, the same verdict would have been returned, had it been collected upon the famous principle of Themistocles, that he should be reputed the first whom the greatest number of rival voices had pronounced to be the second.

Thomas De Quincey

Bibliography

Works: Latin folio, Rome, 1469; Venice, 1471; Florence, 1514; London, 1585. *De Bello Gallico*, Esslingen (?), 1473. Translations by John Tiptoft, Earl of Worcester (John Rastell), of Julius Caesar's Commentaries "newly translated into Englyshe ... as much as concerneth thys realme of England" — 1530 folio; by Arthur Goldinge, *The Eyght Bookes of C. Julius Caesar*, London, 1563, 1565, 1578, 1590; by Chapman, London, 1604 folio; by Clem. Edmonds, London, 1609; the same, with Hirtius, 1655, 1670, 1695 folio with commendatory verses by Camden, Daniel, and Ben Johnson (sic). Works: Translated by W. Duncan, 1753, 1755; by M. Bladen, 8th ed., 1770; MacDevitt, Bohn's Library, 1848. *De Bello Gallico*, translated by R. Mongan, Dublin, 1850; by J.B. Owgan and C.W. Bateman, 1882. *Caesar's Commentaries on the Gallic War*, translated by T. Rice Holmes, London, 1908 (see also Holmes' *Caesar's Conquest of Gaul*, 1911). *Caesar's Gallic War*, translated by Rev. F.P. Long, Oxford, 1911; Books IV. and V. translated by C.H. Prichard, Cambridge, 1912. For Latin text of *De Bello Gallico* see Bell's Illustrated Classical Series; Dent's Temple Series of Classical Texts, 1902; Macmillan and Co., 1905; and Blackie's Latin Texts, 1905-7.

Contents

The Gallic War

The Civil War

BRITANIA
GERMANIA
BÉLGICA
CELTICA
AQUITANIA
NARBONENSE
GALIA CISALPINA
Siluros
Trinobantes
Tamesa
Cantii
Regni
Dumnones
Usipetes
Sicambrios
Catos
Ubios
Moenus
Menapios
Scaldis
Eburones
Nervios
Aduatuca
Aduáticos
Portius Itius
Mórinos
Atrebates
Nemetocerusa
Viromanduos
Mosa
Ambianos
Samarobriga
Bratuspantium
Cáletes
Veliocases
Belóvacos
Remos
Bibrax
Axona
Noviodunum
Suesones
Durocortorum
Tréveros
Vangíones
Németes
Unelos
Lexovios
Sequana
Lutetia
Meldos
Catalauni
Matrona
Mosella
Mediomátricos
Rhenus
Osismos
Coriosolites
Aremorio
Eburovices
Aulercos
Cenómanos
Redones
Parisios
Meclodunum
Tricasses
Leucos
Tribocos
Véneto
Carnutos
Senones
Cenabum
Agedincum
Lingones
Latobicos
Tulingios
Andes
Alesia
Sécuanos
Námnetes
Liger
Noviodunum
Admagetobri(g)a
Vesontio
Ráuracos
Helvecios
Túronos
Heduos
Bituriges
Avaricum
Noviodunum
Dubis
Pictones
Cubos
Decetia
Bibracte
Cabilionum
Jura
Tigurini
Elaver
Lugudunum
Ambarros
Sántonos
Lemovices
Gergovia
Arvernos
Segusiavos
Vienna
Alóbroges
Alpes
Bituriges
Burdigala
Petrocorios
Duranius
Velavios
Boii
Vasates
Nitióbroges
Cadurcos
Oltis
Gábalos
Cevenas
Helvios
Rhodanus
Voconcios
Tannis
Rutenos
Druentia
Tarbelos
Sociates
Auscos
Volcas
Tolosa
Arecómicos
Arelate
Aquae Sextiae
Salluvi/Salves
Massilia
Bigerriones
Garumna
Tectósages
Narbo
Ruscino
Sordones
Pirineos

The Gallic War

Book I

I. — All Gaul is divided into three parts, one of which the Belgae inhabit, the Aquitani another, those who in their own language are called Celts, in ours Gauls, the third. All these differ from each other in language, customs and laws. The river Garonne separates the Gauls from the Aquitani; the Marne and the Seine separate them from the Belgae. Of all these, the Belgae are the bravest, because they are farthest from the civilisation and refinement of [our] Province, and merchants least frequently resort to them and import those things which tend to effeminate the mind; and they are the nearest to the Germans, who dwell beyond the Rhine, with whom they are continually waging war; for which reason the Helvetii also surpass the rest of the Gauls in valour, as they contend with the Germans in almost daily battles, when they either repel them from their own territories, or themselves wage war on their frontiers. One part of these, which it has been said that the Gauls occupy, takes its beginning at the river Rhone: it is bounded by the river Garonne, the ocean, and the territories of the Belgae: it borders, too, on the side of the Sequani and the Helvetii, upon the river Rhine, and stretches towards the north. The Belgae rise from the extreme frontier of Gaul, extend to the lower part of the river Rhine; and look towards the north and the rising sun. Aquitania extends from the river Garonne to the Pyrenaean mountains and to that part of the ocean which is near Spain: it looks between the setting of the sun and the north star.

II. — Among the Helvetii, Orgetorix was by far the most distinguished and wealthy. He, when Marcus Messala and Marcus Piso were consuls, incited by lust of sovereignty, formed a conspiracy among the nobility, and persuaded the people to go forth from their territories with all their possessions, [saying] that it would be very easy, since they excelled all in valour, to acquire the supremacy of the whole of Gaul. To this he the more easily persuaded them, because the Helvetii are confined on every side by the nature of their situation; on one side by the Rhine, a very broad and deep river, which separates the Helvetian territory from the Germans; on a second side by the Jura, a very high mountain which is [situated] between the Sequani and the Helvetii; on a third by the Lake of Geneva, and by the river Rhone, which separates our Province from the Helvetii. From these circumstances it resulted that they could range less widely, and could less easily make war upon their neighbours; for which reason men fond of war [as they were] were affected with great regret. They thought, that considering the extent of their population, and their renown for warfare and bravery, they had but narrow limits, although they extended in length 240, and in breadth 180 [Roman] miles.

III. — Induced by these considerations, and influenced by the authority of Orgetorix, they determined to provide such things as were necessary for their expedition — to buy up as great a number as possible of beasts of burden and waggons — to make their sowings as large as possible, so that on their march plenty of corn might be in store — and to establish peace and friendship with the neighbouring states. They reckoned that a term of two years would be sufficient for them to execute their designs; they fix by decree their departure for the third year. Orgetorix is chosen to complete these arrangements. He took upon himself the office of ambassador to the states: on this journey he persuades Casticus, the son of Catamantaledes (one of the Sequani, whose father had possessed the sovereignty among the people for many years, and had been styled *"friend"* by the senate of the Roman people), to seize upon the sovereignty in his own state, which his father had held before him, and he likewise persuades Dumnorix, an Aeduan, the brother of Divitiacus, who at that time possessed the chief authority in the state, and was exceedingly beloved by the people, to attempt the same, and gives him his daughter in marriage. He proves to them that to accomplish their attempts was a thing very easy to be done, because he himself would obtain the government of his own state; that there was no doubt that the Helvetii were the most powerful of the whole of Gaul; he assures them that he will, with his own forces and his own army, acquire the sovereignty for them. Incited by this speech, they give a pledge and oath to one another, and hope that, when they have seized the sovereignty, they will, by means of the three most powerful and valiant nations, be enabled to obtain possession of the whole of Gaul.

IV. — When this scheme was disclosed to the Helvetii by informers, they, according to their custom, compelled Orgetorix to plead his cause in chains; it was the law that the penalty of being burned by fire should await him if condemned. On the day appointed for the pleading of his cause, Orgetorix drew together from all quarters to the court all his vassals to the number of ten thousand persons; and led together to the same place, and all his dependants and debtor-bondsmen, of whom he had a great number; by means of these he rescued himself from [the necessity of] pleading his cause. While the state, incensed at this act, was endeavouring to assert its right by arms, and the magistrates were mustering a large body of men from the country, Orgetorix died; and there is not wanting a suspicion, as the Helvetii think, of his having committed suicide.

V. — After his death, the Helvetii nevertheless attempt to do that which they had resolved on, namely, to go forth from their territories. When they thought that they were at length prepared for this undertaking, they set fire to all their towns, in number about twelve — to their villages about four hundred — and to the private dwellings that remained; they burn up all the corn, except what they intend to carry with them; that after destroying the hope of a return home, they might be the more ready for undergoing all dangers. They order every one to carry forth from home for himself provisions for three months, ready ground. They persuade the Rauraci, and the Tulingi, and the Latobrigi, their neighbours, to adopt the same plan, and after burning down their towns and villages, to set out with them: and they admit to their party and unite to themselves as confederates the Boii, who had dwelt on the other side of the Rhine, and had crossed over into the Norican territory, and assaulted Noreia.

VI. — There were in all two routes by which they could go forth from their country — one through the Sequani, narrow and difficult, between Mount Jura and the river Rhone (by which scarcely one waggon at a time could be led; there was, moreover, a very high mountain overhanging, so that a very few might easily intercept them); the other, through our Province, much easier and freer from obstacles, because the Rhone flows between the boundaries of the Helvetii and those of the Allobroges, who had lately been subdued, and is in some places crossed by a ford. The furthest town of the Allobroges, and the nearest to the territories of the Helvetii, is Geneva. From this town a bridge extends to the Helvetii. They thought that they should either persuade the Allobroges, because they did not seem as yet well-affected towards the Roman people, or compel them by force to allow them to pass through their territories. Having provided everything for the expedition, they appoint a day on which they should all meet on the bank of the Rhone. This day was the fifth before the kalends of April [*i.e.* the 28th of March], in the consulship of Lucius Piso and Aulus Gabinius [B.C. 58].

VII. — When it was reported to Caesar that they were attempting to make their route through our Province, he hastens to set out from the city, and, by as great marches as he can, proceeds to Further Gaul, and arrives at Geneva. He orders the whole Province [to furnish] as great a number of soldiers as possible, as there was in all only one legion in Further Gaul: he orders the bridge at Geneva to be broken down. When the Helvetii are apprised of his arrival, they send to him, as ambassadors, the most illustrious men of their state (in which embassy Numeius and Verudoctius held the chief place), to say "that it was their intention to march through the Province without doing any harm, because they had" [according to their own representations] "no other route: — that they requested they might be allowed to do so with his consent." Caesar, inasmuch as he kept in remembrance that Lucius Cassius, the consul, had been slain, and his army routed and made to pass under the yoke by the Helvetii, did not think that [their request] ought to be granted; nor was he of opinion that men of hostile disposition, if an opportunity of marching through the

Province were given them, would abstain from outrage and mischief. Yet, in order that a period might intervene, until the soldiers whom he had ordered [to be furnished] should assemble, he replied to the ambassadors, that he would take time to deliberate; if they wanted anything, they might return on the day before the ides of April [on April 12th].

VIII. — Meanwhile, with the legion which he had with him and the soldiers who had assembled from the Province, he carries along for nineteen [Roman, not quite eighteen English] miles a wall, to the height of sixteen feet, and a trench, from the lake of Geneva, which flows into the river Rhone, to Mount Jura, which separates the territories of the Sequani from those of the Helvetii. When that work was finished, he distributes garrisons, and closely fortifies redoubts, in order that he may the more easily intercept them, if they should attempt to cross over against his will. When the day which he had appointed with the ambassadors came, and they returned to him, he says that he cannot, consistently with the custom and precedent of the Roman people, grant any one a passage through the Province; and he gives them to understand that, if they should attempt to use violence, he would oppose them. The Helvetii, disappointed in this hope, tried if they could force a passage (some by means of a bridge of boats and numerous rafts constructed for the purpose; others, by the fords of the Rhone, where the depth of the river was least, sometimes by day, but more frequently by night), but being kept at bay by the strength of our works, and by the concourse of the soldiers, and by the missiles, they desisted from this attempt.

IX. — There was left one way, [namely] through the Sequani, by which, on account of its narrowness, they could not pass without the consent of the Sequani. As they could not of themselves prevail on them, they send ambassadors to Dumnorix the Aeduan, that through his intercession they might obtain their request from the Sequani. Dumnorix, by his popularity and liberality, had great influence among the Sequani, and was friendly to the Helvetii, because out of that state he had married the daughter of Orgetorix; and, incited by lust of sovereignty, was anxious for a revolution, and wished to have as many states as possible attached to him by his kindness towards them. He, therefore, undertakes the affair, and prevails upon the Sequani to allow the Helvetii to march through their territories, and arranges that they should give hostages to each other — the Sequani not to obstruct the Helvetii in their march — the Helvetii, to pass without mischief and outrage.

X. — It is again told Caesar that the Helvetii intend to march through the country of the Sequani and the Aedui into the territories of the Santones, which are not far distant from those boundaries of the Tolosates, which [viz. Tolosa, Toulouse] is a state in the Province. If this took place, he saw that it would be attended with great danger to the Province to have warlike men, enemies of the Roman people, bordering upon an open and very fertile tract of country. For these reasons he appointed Titus Labienus, his lieutenant, to the command of the fortification which he had made. He himself proceeds to Italy by forced marches, and there levies two legions, and leads out from winter-quarters three which were wintering around Aquileia, and with these five legions marches rapidly by the nearest route across the Alps into Further Gaul. Here the Centrones and the Graioceli and the Caturiges, having taken possession of the higher parts, attempt to obstruct the army in their march. After having routed these in several battles, he arrives in the territories of the Vocontii in the Further Province on the seventh day from Ocelum, which is the most remote town of the Hither Province; thence he leads his army into the country of the Allobroges, and from the Allobroges to the Segusiani. These people are the first beyond the Province on the opposite side of the Rhone.

XI. — The Helvetii had by this time led their forces over through the narrow defile and the territories of the Sequani, and had arrived at the territories of the Aedui, and were ravaging their lands. The Aedui, as they could not defend themselves and their possessions against them, send ambassadors to Caesar to ask assistance, [pleading] that they had at all times so well deserved of the Roman people, that their fields ought not to have been laid waste — their children carried off into slavery — their towns stormed, almost within sight of our army. At the same time the Ambarri, the friends and kinsmen of the Aedui, apprise Caesar that it was not easy for them, now that their fields had been devastated, to ward off the violence of the enemy from their towns: the Allobroges likewise, who had villages and possessions on the other side of the Rhone, betake themselves in flight to Caesar and assure him that they had nothing remaining, except the soil of their land. Caesar, induced by these circumstances, decides that he ought not to wait until the Helvetii, after destroying all the property of his allies, should arrive among the Santones.

XII. — There is a river [called] the Saone, which flows through the territories of the Aedui and Sequani into the Rhone with such incredible slowness, that it cannot be determined by the eye in which direction it flows. This the Helvetii were crossing by rafts and boats joined together. When Caesar was informed by spies that the Helvetii had already conveyed three parts of their forces across that river, but that the fourth part was left behind on this side of the Saone, he set out from the camp with three legions during the third watch, and came up with that division which had not yet crossed the river. Attacking them, encumbered with baggage, and not expecting him, he cut to pieces a great part of them; the rest betook themselves to flight, and concealed themselves in the nearest woods. That canton [which was cut down] was called the Tigurine; for the whole Helvetian state is divided into four cantons. This single canton having left their country, within the recollection of our fathers, had slain Lucius Cassius the consul, and had made his army pass under the yoke [B.C. 107]. Thus, whether by chance, or by the design of the immortal gods, that part of the Helvetian state which had brought a signal calamity upon the Roman people was the first to pay the penalty. In this Caesar avenged not only the public, but also his own personal wrongs, because the Tigurini had slain Lucius Piso the lieutenant [of Cassius], the grandfather of Lucius Calpurnius Piso, his [Caesar's] father-in-law, in the same battle as Cassius himself.

XIII. — This battle ended, that he might be able to come up with the remaining forces of the Helvetii, he procures a bridge to be made across the Saone, and thus leads his army over. The Helvetii, confused by his sudden arrival, when they found that he had effected in one day what they themselves had with the utmost difficulty accomplished in twenty, namely, the crossing of the river, send ambassadors to him; at the head of which embassy was Divico, who had been commander of the Helvetii in the war against Cassius. He thus treats with Caesar: — that, "if the Roman people would make peace with the Helvetii they would go to that part and there remain, where Caesar might appoint and desire them to be; but if he should persist in persecuting them with war, that he ought to remember both the ancient disgrace of the Roman people and the characteristic valour of the Helvetii. As to his having attacked one canton by surprise, [at a time] when those who had crossed the river could not bring assistance to their friends, that he ought not on that account to ascribe very much to his own valour, or despise them; that they had so learned from their sires and ancestors, as to rely more on valour than on artifice or stratagem. Wherefore let him not bring it to pass that the place, where they were standing, should acquire a name, from the disaster of the Roman people and the destruction of their army or transmit the remembrance [of such an event to posterity]."

XIV. — To these words Caesar thus replied: — that "on that very account he felt less hesitation, because he kept in remembrance those circumstances which the Helvetian ambassadors had mentioned, and that he felt the more indignant at them, in proportion as they had happened undeservedly to the Roman people: for if they had been conscious of having done any wrong it would not have been difficult to be on their guard, but for that very reason had they been deceived, because neither were they aware that any offence had been given by them, on account of which they should be afraid, nor did they think that they ought to be afraid without cause. But even if he were willing to forget their former outrage, could he also lay aside the remembrance of the late wrongs, in that they had against his will attempted a route through the Province by force, in that they had molested the Aedui, the Ambarri, and the Allobroges? That as to their so insolently boasting of their victory, and as to their being astonished that they had so long committed their outrages with impunity, [both these things] tended to the same point; for the immortal gods are wont to allow those persons whom they wish to punish for their guilt sometimes a greater prosperity and longer impunity, in order that they may suffer the more severely from a reverse of circumstances. Although these things are so, yet, if hostages were to be given him by them in order that he may be assured they will do what they promise, and provided they will give satisfaction to the Aedui for the outrages which they had committed against them and their allies, and likewise to the Allobroges, he [Caesar] will make peace with them." Divico replied, that "the Helvetii had been so trained by their ancestors that they were accustomed to receive, not to give, hostages; of that fact the Roman people were witness." Having given this reply, he withdrew.

XV. — On the following day they move their camp from that place; Caesar does the same, and sends forward all his cavalry, to the number of four thousand (which he had drawn together from all parts of the Province and from the Aedui and their allies), to observe towards what parts the enemy are directing their march. These, having too eagerly pursued the enemy's rear, come to a battle with the cavalry of the Helvetii in a disadvantageous place, and a few of our men fall. The Helvetii, elated with this battle because they had with five hundred horse repulsed so large a body of horse, began to face us more boldly, sometimes too from their rear to provoke our men by an attack. Caesar [however] restrained his men from battle, deeming it sufficient for the present to prevent the enemy from rapine, forage, and depredation. They marched for about fifteen days in such a manner that there was not more than five or six miles between the enemy's rear and our van.

XVI. — Meanwhile, Caesar kept daily importuning the Aedui for the corn which they had promised in the name of their state; for, in consequence of the coldness (Gaul being, as before said, situated towards the north), not only was the corn in the fields not ripe, but there was not in store a sufficiently large quantity even of fodder: besides he was unable to use the corn which he had conveyed in ships up the river Saone, because the Helvetii, from whom he was unwilling to retire, had diverted their march from the Saone. The Aedui kept deferring from day to day, and saying that it was being "collected — brought in — on the road." When he saw that he was put off too long, and that the day was close at hand on which he ought to serve out the corn to his soldiers, — having called together their chiefs, of whom he had a great number in his camp, among them Divitiacus, and Liscus who was invested with the chief magistracy (whom the Aedui style the Vergobretus, and who is elected annually, and has power of life and death over his countrymen), he severely reprimands them, because he is not assisted by them on so urgent an occasion, when the enemy were so close at hand, and when [corn] could neither be bought nor taken from the fields, particularly as, in a great measure urged by their prayers, he had undertaken the war; much more bitterly, therefore, does he complain of his being forsaken.

XVII. — Then at length Liscus, moved by Caesar's speech, discloses what he had hitherto kept secret: — that "there are some whose influence with the people is very great, who, though private men, have more power than the magistrates themselves: that these by seditious and violent language are deterring the populace from contributing the corn which they ought to supply; [by telling them] that, if they cannot any longer retain the supremacy of Gaul, it were better to submit to the government of Gauls than of Romans, nor ought they to doubt that, if the Romans should overpower the Helvetii, they would wrest their freedom from the Aedui together with the remainder of Gaul. By these very men [said he] are our plans, and whatever is done in the camp, disclosed to the enemy; that they could not be restrained by *him*: nay more, he was well aware that, though compelled by necessity, he had disclosed the matter to Caesar, at how great a risk he had done it; and for that reason, he had been silent as long as he could."

XVIII. — Caesar perceived that, by this speech of Liscus, Dumnorix, the brother of Divitiacus, was indicated; but, as he was unwilling that these matters should be discussed while so many were present, he speedily dismisses the council, but detains Liscus: he inquires from him when alone, about those things which he had said in the meeting. He [Liscus] speaks more unreservedly and boldly. He [Caesar] makes inquiries on the same points privately of others, and discovers that it is all true; that "Dumnorix is the person, a man of the highest daring, in great favour with the people on account of his liberality, a man eager for a revolution: that for a great many years he has been in the habit of contracting for the customs and all the other taxes of the Aedui at a small cost, because when *he* bids, no one dares to bid against him. By these means he has both increased his own private property and amassed great means for giving largesses; that he maintains constantly at his own expense and keeps about his own person a great number of cavalry, and that not only at home, but even among the neighbouring states, he has great influence, and for the sake of strengthening this influence has given his mother in marriage among the Bituriges to a man the most noble and most influential there; that he has himself taken a wife from among the Helvetii, and has given his sister by the mother's side and his female relations in marriage into other states; that he favours and wishes well to the Helvetii on account of this connection; and that he hates Caesar and the Romans, on his own account, because by their arrival his power was weakened, and his brother,

Divitiacus, restored to his former position of influence and dignity: that, if anything should happen to the Romans, he entertains the highest hope of gaining the sovereignty by means of the Helvetii, but that under the government of the Roman people he despairs not only of royalty but even of that influence which he already has." Caesar discovered too, on inquiring into the unsuccessful cavalry engagement which had taken place a few days before, that the commencement of that flight had been made by Dumnorix and his cavalry (for Dumnorix was in command of the cavalry which the Aedui had sent for aid to Caesar); that by their flight the rest of the cavalry was dismayed.

XIX. — After learning these circumstances, since to these suspicions the most unequivocal facts were added, viz., that he had led the Helvetii through the territories of the Sequani; that he had provided that hostages should be mutually given; that he had done all these things, not only without any orders of his [Caesar's] and of his own state's, but even without their [the Aedui] knowing anything of it themselves; that he [Dumnorix] was reprimanded by the [chief] magistrate of the Aedui; he [Caesar] considered that there was sufficient reason why he should either punish him himself, or order the state to do so. One thing [however] stood in the way of all this — that he had learned by experience his brother Divitiacus's very high regard for the Roman people, his great affection towards him, his distinguished faithfulness, justice, and moderation; for he was afraid lest by the punishment of this man, he should hurt the feelings of Divitiacus. Therefore, before he attempted anything, he orders Divitiacus to be summoned to him, and when the ordinary interpreters had been withdrawn, converses with him through Caius Valerius Procillus, chief of the province of Gaul, an intimate friend of his, in whom he reposed the highest confidence in everything; at the same time he reminds him of what was said about Dumnorix in the council of the Gauls, when he himself was present, and shows what each had said of him privately in his [Caesar's] own presence; he begs and exhorts him, that, without offence to his feelings, he may either himself pass judgment on him [Dumnorix] after trying the case, or else order the [Aeduan] state to do so.

XX. — Divitiacus, embracing Caesar, begins to implore him, with many tears, that "he would not pass any very severe sentence upon his brother; saying, that he knows that those [charges] are true, and that nobody suffered more pain on that account than he himself did; for when he himself could effect a very great deal by his influence at home and in the rest of Gaul, and he [Dumnorix] very little on account of his youth, the latter had become powerful through his means, which power and strength he used not only to the lessening of his [Divitiacus] popularity, but almost to his ruin; that he, however, was influenced both by fraternal affection and by public opinion. But if anything very severe from Caesar should befall him [Dumnorix], no one would think that it had been done without his consent, since he himself held such a place in Caesar's friendship; from which circumstance it would arise that the affections of the whole of Gaul would be estranged from him." As he was with tears begging these things of Caesar in many words, Caesar takes his right hand, and, comforting him, begs him to make an end of entreating, and assures him that his regard for him is so great that he forgives both the injuries of the republic and his private wrongs, at his desire and prayers. He summons Dumnorix to him; he brings in his brother; he points out what he censures in him; he lays before him what he of himself perceives, and what the state complains of; he warns him for the future to avoid all grounds of suspicion; he says that he pardons the past, for the sake of his brother, Divitiacus. He sets spies over Dumnorix that he may be able to know what he does, and with whom he communicates.

XXI. — Being on the same day informed by his scouts that the enemy had encamped at the foot of a mountain eight miles from his own camp, he sent persons to ascertain what the nature of the mountain was, and of what kind the ascent on every side. Word was brought back that it was easy. During the third watch he orders Titus Labienus, his lieutenant with praetorian powers, to ascend to the highest ridge of the mountain with two legions, and with those as guides who had examined the road; he explains what his plan is. He himself during the fourth watch, hastens to them by the same route by which the enemy had gone, and sends on all the cavalry before him. Publius Considius, who was reputed to be very experienced in military affairs, and had been in the army of Lucius Sulla, and afterwards in that of Marcus Crassus, is sent forward with the scouts.

XXII. — At day-break, when the summit of the mountain was in the possession of Titus Labienus, and he himself was not further off than a mile and half from the enemy's camp, nor, as he afterwards ascertained from the captives, had either his arrival or that of Labienus been discovered; Considius, with his horse at full gallop, comes up to him — says that the mountain which he [Caesar] wished should be seized by Labienus, is in possession of the enemy; that he has discovered this by the Gallic arms and ensigns. Caesar leads off his forces to the next hill: [and] draws them up in battle-order. Labienus, as he had been ordered by Caesar not to come to an engagement unless [Caesar's] own forces were seen near the enemy's camp, that the attack upon the enemy might be made on every side at the same time, was, after having taken possession of the mountain, waiting for our men, and refraining from battle. When, at length, the day was far advanced, Caesar learned through spies that the mountain was in possession of his own men, and that the Helvetii had moved their camp, and that Considius, struck with fear, had reported to him, as seen, that which he had not seen. On that day he follows the enemy at his usual distance, and pitches his camp three miles from theirs.

XXIII. — The next day (as there remained in all only two days' space [to the time] when he must serve out the corn to his army, and as he was not more than eighteen miles from Bibracte, by far the largest and best-stored town of the Aedui) he thought that he ought to provide for a supply of corn; and diverted his march from the Helvetii, and advanced rapidly to Bibracte. This circumstance is reported to the enemy by some deserters from Lucius Aemilius, a captain of the Gallic horse. The Helvetii, either because they thought that the Romans, struck with terror, were retreating from them, the more so, as the day before, though they had seized on the higher grounds, they had not joined battle; or because they flattered themselves that they might be cut off from the provisions, altering their plan and changing their route, began to pursue and to annoy our men in the rear.

XXIV. — Caesar, when he observes this, draws off his forces to the next hill, and sent the cavalry to sustain the attack of the enemy. He himself, meanwhile, drew up on the middle of the hill a triple line of his four veteran legions in such a manner that he placed above him on the very summit the two legions which he had lately levied in Hither Gaul, and all the auxiliaries; and he ordered that the whole mountain should be covered with men, and that meanwhile the baggage should be brought together into one place, and the position be protected by those who were posted in the upper line. The Helvetii, having followed with all their waggons, collected their baggage into one place: they themselves, after having repulsed our cavalry and formed a phalanx, advanced up to our front line in very close order.

XXV. — Caesar, having removed out of sight first his own horse, then those of all, that he might make the danger of all equal, and do away with the hope of flight, after encouraging his men, joined battle. His soldiers, hurling their javelins from the higher ground, easily broke the enemy's phalanx. That being dispersed, they made a charge on them with drawn swords. It was a great hindrance to the Gauls in fighting, that, when several of their bucklers had been by one stroke of the (Roman) javelins pierced through and pinned fast together, as the point of the iron had bent itself, they could neither pluck it out, nor, with their left hand entangled, fight with sufficient ease; so that many, after having long tossed their arm about, chose rather to cast away the buckler from their hand, and to fight with their person unprotected. At length, worn out with wounds, they began to give way, and as there was in the neighbourhood a mountain about a mile off, to betake themselves thither. When the mountain had been gained, and our men were advancing up, the Boii and Tulingi, who with about 15,000 men closed the enemy's line of march and served as a guard to their rear, having assailed our men on the exposed flank as they advanced [prepared] to surround them; upon seeing which, the Helvetii, who had betaken themselves to the mountain, began to press on again and renew the battle. The Romans having faced about, advanced to the attack in two divisions; the first and second line to withstand those who had been defeated and driven off the field; the third to receive those who were just arriving.

XXVI. — Thus was the contest long and vigorously carried on with doubtful success. When they could no longer withstand the attacks of our men, the one division, as they had begun to do, betook themselves to the mountain; the other repaired to their baggage and waggons. For during the whole of this battle, although the fight lasted from the seventh hour [*i.e.* 12 (noon) — 1 P.M.] to eventide, no one could see an enemy with his back turned. The fight was carried on also at the baggage till late in the night, for they had set waggons in the way as a rampart, and from the higher ground kept throwing weapons upon our men, as they came on, and some from between the waggons and the wheels kept darting their lances and javelins from beneath, and wounding our men. After the fight had lasted some time, our men gained possession of their baggage and camp. There the daughter and one of the sons of Orgetorix were taken. After that battle about 130,000 men [of the enemy] remained alive, who marched incessantly during the whole of that night; and after a march discontinued for no part of the night, arrived in the territories of the Lingones on the fourth day, whilst our men, having stopped for three days, both on account of the wounds of the soldiers and the burial of the slain, had not been able to follow them. Caesar sent letters and messengers to the Lingones [with orders] that they should not assist them with corn or with anything else; for that if they should assist them, he would regard them in the same light as the Helvetii. After the three days' interval he began to follow them himself with all his forces.

XXVII. — The Helvetii, compelled by the want of everything, sent ambassadors to him about a surrender. When these had met him in the way and had thrown themselves at his feet, and speaking in suppliant tone had with tears sued for peace, and [when] he had ordered them to await his arrival, in the place where they then were, they obeyed his commands. When Caesar arrived at that place, he demanded hostages, their arms, and the slaves who had deserted to them. Whilst those things are being sought for and got together, after a night's interval, about 6000 men of that canton which is called the Verbigene, whether terrified by fear, lest, after delivering up their arms, they should suffer punishment, or else induced by the hope of safety, because they supposed that, amid so vast a multitude of those who had surrendered themselves, *their* flight might either be concealed or entirely overlooked, having at night-fall departed out of the camp of the Helvetii, hastened to the Rhine and the territories of the Germans.

XXVIII. — But when Caesar discovered this, he commanded those through whose territories they had gone, to seek them out and to bring them back again, if they meant to be acquitted before him; and considered them, when brought back, in the light of enemies; he admitted all the rest to a surrender, upon their delivering up the hostages, arms, and deserters. He ordered the Helvetii, the Tulingi, and the Latobrigi to return to their territories from which they had come, and as there was at home nothing whereby they might support their hunger, all the productions of the earth having been destroyed, he commanded the Allobroges to let them have a plentiful supply of corn; and ordered them to rebuild the towns and villages which they had burnt. This he did, chiefly on this account, because he was unwilling that the country, from which the Helvetii had departed, should be untenanted, lest the Germans, who dwell on the other side of the Rhine, should, on account of the excellence of the lands, cross over from their own territories into those of the Helvetii, and become borderers upon the province of Gaul and the Allobroges. He granted the petition of the Aedui, that they might settle the Boii, in their own (*i.e.* in the Aeduan) territories, as these were known to be of distinguished valour to whom they gave lands, and whom they afterwards admitted to the same state of rights and freedom as themselves.

XXIX. — In the camp of the Helvetii, lists were found, drawn up in Greek characters, and were brought to Caesar, in which an estimate had been drawn up, name by name, of the number which had gone forth from their country of those who were able to bear arms; and likewise the boys, the old men, and the women, separately. Of all which items the total was:—

Of the *Helvetii* [lit. of the heads of the Helvetii] 263,000
Of the *Tulingi* 36,000
Of the *Latobrigi* 14,000
Of the *Rauraci* 23,000
Of the *Boii* 32,000

The sum of all amounted to 368,000

Out of these, such as could bear arms [amounted] to about 92,000. When the census of those who returned home was taken, as Caesar had commanded, the number was found to be 110,000.

XXX. — When the war with the Helvetii was concluded, ambassadors from almost all parts of Gaul, the chiefs of states, assembled to congratulate Caesar, [saying] that they were well aware, that, although he had taken vengeance on the Helvetii in war, for the old wrongs done by them to the Roman people, yet that circumstance had happened no less to the benefit of the land of Gaul than of the Roman people, because the Helvetii, while their affairs were most flourishing, had quitted their country with the design of making war upon the whole of Gaul, and seizing the government of it, and selecting, out of a great abundance, that spot for an abode which they should judge to be the most convenient and most productive of all Gaul, and hold the rest of the states as tributaries. They requested that they might be allowed to proclaim an assembly of the whole of Gaul for a particular day, and to do that with Caesar's permission, [stating] that they had some things which, with the general consent, they wished to ask of him. This request having been granted, they appointed a day for the assembly, and ordained by an oath with each other, that no one should disclose [their deliberations] except those to whom this [office] should be assigned by the general assembly.

XXXI. — When that assembly was dismissed, the same chiefs of states, who had before been to Caesar, returned, and asked that they might be allowed to treat with him privately (in secret) concerning the safety of themselves and of all. That request having been obtained, they all threw themselves in tears at Caesar's feet, [saying] that they no less begged and earnestly desired that what they might say should not be disclosed than that they might obtain those things which they wished for; inasmuch as they saw that, if a disclosure were made, they should be put to the greatest tortures. For these Divitiacus the Aeduan spoke and told him: — "That there were two parties in the whole of Gaul: that the Aedui stood at the head of one of these, the Arverni of the other. After these had been violently struggling with one another for the superiority for many years, it came to pass that the Germans were called in for hire by the Arverni and the Sequani. That about 15,000 of them [*i.e.* of the Germans] had at first crossed the Rhine: but after that these wild and savage men had become enamoured of the lands and the refinement and the abundance of the Gauls, more were brought over, that there were now as many as 120,000 of them in Gaul: that with these the Aedui and their dependants had repeatedly struggled in arms, that they had been routed and had sustained a great calamity — had lost all their nobility, all their senate, all their cavalry. And that broken by such engagements and calamities, although they had formerly been very powerful in Gaul, both from their own valour and from the Roman people's hospitality and friendship, they were now compelled to give the chief nobles of their state as hostages to the Sequani, and to bind their state by an oath, that they would neither demand hostages in return, nor supplicate aid from the Roman people, nor refuse to be for ever under their sway and empire. That he was the only one out of all the state of the Aedui who could not be prevailed upon to take the oath or to give his children as hostages. On that account he had fled from his state and had gone to the senate at Rome to beseech aid, as he alone was bound neither by oath nor hostages. But a worse thing had befallen the victorious Sequani than the vanquished Aedui, for

Ariovistus, the king of the Germans, had settled in their territories, and had seized upon a third of their land, which was the best in the whole of Gaul, and was now ordering them to depart from another third part, because a few months previously 24,000 men of the Harudes had come to him, for whom room and settlements must be provided. The consequence would be, that in a few years they would all be driven from the territories of Gaul, and all the Germans would cross the Rhine; for neither must the land of Gaul be compared with the land of the Germans, nor must the habit of living of the latter be put on a level with that of the former. Moreover, [as for] Ariovistus, no sooner did he defeat the forces of the Gauls in a battle, which took place at Magetobria, than [he began] to lord it haughtily and cruelly, to demand as hostages the children of all the principal nobles, and wreak on them every kind of cruelty, if everything was not done at his nod or pleasure; that he was a savage, passionate, and reckless man, and that his commands could no longer be borne. Unless there was some aid in Caesar and the Roman people, the Gauls must all do the same thing that the Helvetii had done, [viz.] emigrate from their country, and seek another dwelling place, other settlements remote from the Germans, and try whatever fortune may fall to their lot. If these things were to be disclosed to Ariovistus, [Divitiacus adds] that he doubts not that he would inflict the most severe punishment on all the hostages who are in his possession, [and says] that Caesar could, either by his own influence and by that of his army, or by his late victory, or by name of the Roman people, intimidate him, so as to prevent a greater number of Germans being brought over the Rhine, and could protect all Gaul from the outrages of Ariovistus."

XXXII.—When this speech had been delivered by Divitiacus, all who were present began with loud lamentation to entreat assistance of Caesar. Caesar noticed that the Sequani were the only people of all who did none of those things which the others did, but, with their heads bowed down, gazed on the earth in sadness. Wondering what was the reason of this conduct, he inquired of themselves. No reply did the Sequani make, but silently continued in the same sadness. When he had repeatedly inquired of them and could not elicit any answer at all, the same Divitiacus the Aeduan answered, that—"the lot of the Sequani was more wretched and grievous than that of the rest, on this account, because they alone durst not even in secret complain or supplicate aid; and shuddered at the cruelty of Ariovistus [even when] absent, just as if he were present; for, to the rest, despite of everything, there was an opportunity of flight given; but all tortures must be endured by the Sequani, who had admitted Ariovistus within their territories, and whose towns were all in his power."

XXXIII.— Caesar, on being informed of these things, cheered the minds of the Gauls with his words, and promised that this affair should be an object of his concern, [saying] that he had great hopes that Ariovistus, induced both by his kindness and his power, would put an end to his oppression. After delivering this speech, he dismissed the assembly; and, besides those statements, many circumstances induced him to think that this affair ought to be considered and taken up by him; especially as he saw that the Aedui, styled [as they had been] repeatedly by the senate "brethren" and "kinsmen," were held in the thraldom and dominion of the Germans, and understood that their hostages were with Ariovistus and the Sequani, which in so mighty an empire [as that] of the Roman people he considered very disgraceful to himself and the republic. That, moreover, the Germans should by degrees become accustomed to cross the Rhine, and that a great body of them should come into Gaul, he saw [would be] dangerous to the Roman people, and judged that wild and savage men would not be likely to restrain themselves, after they had possessed themselves of all Gaul, from going forth into the province and thence marching into Italy (as the Cimbri and Teutones had done before them), particularly as the Rhone [was the sole barrier that] separated the Sequani from our province. Against which events he thought he ought to provide as speedily as possible. Moreover, Ariovistus, for his part, had assumed to himself such pride and arrogance that he was felt to be quite insufferable.

XXXIV. — He therefore determined to send ambassadors to Ariovistus to demand of him to name some intermediate spot for a conference between the two, [saying] that he wished to treat with him on state-business and matters of the highest importance to both of them. To this embassy Ariovistus replied, that if he himself had had need of anything from Caesar, he would have gone to him; and that if Caesar wanted anything from him he ought to come to him. That, besides, neither dare he go without an army into those parts of Gaul which Caesar had possession of, nor could he, without great expense and trouble, draw his army together to one place; that to him, moreover, it appeared strange what business either Caesar or the Roman people at all had in his own Gaul, which he had conquered in war.

XXXV. — When these answers were reported to Caesar, he sends ambassadors to him a second time with this message "Since, after having been treated with so much kindness by himself and the Roman people (as he had in his consulship [B.C. 59] been styled 'king and friend' by the senate), he makes this recompense to [Caesar] himself and the Roman people, [viz.] that when invited to a conference he demurs, and does not think that it concerns him to advise and inform himself about an object of mutual interest, these are the things which he requires of him; first, that he do not any more bring over any body of men across the Rhine into Gaul; in the next place, that he restore the hostages which he has from the Aedui, and grant the Sequani permission to restore to them with his consent those hostages which they have, and that he neither provoke the Aedui by outrage nor make war upon them or their allies; if he would accordingly do this," [Caesar says] that "he himself and the Roman people will entertain a perpetual feeling of favour and friendship towards him; but that if he [Caesar] does not obtain [his desires], that he (forasmuch as in the consulship of Marcus Messala and Marcus Piso [B.C. 61] the senate had decreed that, whoever should have the administration of the province of Gaul should, as far as he could do so consistently with the interests of the republic, protect the Aedui and the other friends of the Roman people) will not overlook the wrongs of the Aedui."

XXXVI. — To this Ariovistus replied, that "the right of war was, that they who had conquered should govern those whom they had conquered, in what manner they pleased; that in that way the Roman people were wont to govern the nations which they had conquered, not according to the dictation of any other, but according to their own discretion. If he for his part did not dictate to the Roman people as to the manner in which they were to exercise their right, he ought not to be obstructed by the Roman people in his right; that the Aedui, inasmuch as they had tried the fortune of war and had engaged in arms and been conquered, had become tributaries to him; that Caesar was doing a great injustice, in that by his arrival he was making his revenues less valuable to him; that he should not restore their hostages to the Aedui, but should not make war wrongfully either upon them or their allies, if they abided by that which had been agreed on, and paid their tribute annually: if they did *not* continue to do that, the Roman people's name of 'brothers' would avail them nought. As to Caesar's threatening him that be would not overlook the wrongs of the Aedui, [he said] that no one had ever entered into a contest with him [Ariovistus] without utter ruin to himself. That Caesar might enter the lists when he chose; he would feel what the invincible Germans, well-trained [as they were] beyond all others to arms, who for fourteen years had not been beneath a roof, could achieve by their valour."

XXXVII. — At the same time that this message was delivered to Caesar, ambassadors came from the Aedui and the Treviri; from the Aedui to complain that the Harudes, who had lately been brought over into Gaul, were ravaging their territories; that they had not been able to purchase peace from Ariovistus, even by giving hostages: and from the Treviri, [to state] that a hundred cantons of the Suevi had encamped on the banks of the Rhine, and were attempting to cross it; that the brothers, Nasuas and Cimberius, headed them. Being greatly alarmed at these things, Caesar thought that he ought to use all despatch, lest, if this new band of Suevi should unite with the old troops of Ariovistus, he [Ariovistus] might be less easily withstood. Having, therefore, as quickly as he could, provided a supply of corn, he hastened to Ariovistus by forced marches.

XXXVIII. — When he had proceeded three days' journey, word was brought to him that Ariovistus was hastening with all his forces to seize on Vesontio, which is the largest town of the Sequani, and had advanced three days' journey from his territories. Caesar thought that he ought to take the greatest precautions lest this should happen, for there was in that town a most ample supply of everything which was serviceable for war; and so fortified was it by the nature of the ground as to afford a great facility for protracting the war, inasmuch as the river Doubs almost surrounds the whole town, as though it were traced round it with a pair of compasses. A mountain of great height shuts in the remaining space, which is not more than 600 feet, where the river leaves a gap, in such a manner that the roots of that mountain extend to the river's bank on either side. A wall thrown around it makes a citadel of this [mountain], and connects it with the town. Hither Caesar hastens by forced marches by night and day, and, after having seized the town, stations a garrison there.

XXXIX. — Whilst he is tarrying a few days at Vesontio, on account of corn and provisions; from the inquiries of our men and the reports of the Gauls and traders (who asserted that the Germans were men of huge stature, of incredible valour and practice in arms, that ofttimes they, on encountering them, could not bear even their countenance, and the fierceness of their eyes) — so great a panic on a sudden seized the whole army, as to discompose the minds and spirits of all in no slight degree. This first arose from the tribunes of the soldiers, the prefects and the rest, who, having followed Caesar from the city [Rome] from motives of friendship, had no great experience in military affairs. And alleging, some of them one reason, some another, which they said made it necessary for them to depart, they requested that by his consent they might be allowed to withdraw; some, influenced by shame, stayed behind in order that they might avoid the suspicion of cowardice. These could neither compose their countenance, nor even sometimes check their tears: but hidden in their tents, either bewailed their fate, or deplored with their comrades the general danger. Wills were sealed universally throughout the whole camp. By the expressions and cowardice of these men, even those who possessed great experience in the camp, both soldiers and centurions, and those [the decurions] who were in command of the cavalry, were gradually disconcerted. Such of them as wished to be considered less alarmed, said that they did not dread the enemy, but feared the narrowness of the roads and the vastness of the forests which lay between them and Ariovistus, or else that the supplies could not be brought up readily enough. Some even declared to Caesar that when he gave orders for the camp to be moved and the troops to advance, the soldiers would not be obedient to the command, nor advance in consequence of their fear.

XL. — When Caesar observed these things, having called a council, and summoned to it the centurions of all the companies, he severely reprimanded them, "particularly for supposing that it belonged to them to inquire or conjecture, either in what direction they were marching, or with what object. That Ariovistus, during his [Caesar's] consulship, had most anxiously sought after the friendship of the Roman people; why should any one judge that he would so rashly depart from his duty? He for his part was persuaded that, when his demands were known and the fairness of the terms considered, he would reject neither his nor the Roman people's favour. But even if, driven on by rage and madness, he should make war upon them, what after all were they afraid of? — or why should they despair either of their own valour or of his zeal? Of that enemy a trial had been made within our fathers' recollection, when, on the defeat of the Cimbri and Teutones by Caius Marius, the army was regarded as having deserved no less praise than their commander himself. It had been made lately, too, in Italy; during the rebellion of the slaves, whom, however, the experience and training which they had received from us, assisted in some respect. From which a judgment might be formed of the advantages which resolution carries with it, — inasmuch as those whom for some time they had groundlessly dreaded when unarmed, they had afterwards vanquished, when well armed and flushed with success. In short, that these were the same men whom the Helvetii, in frequent encounters, not only in their own territories, but also in theirs [the German], have generally vanquished, and yet cannot have been a match for our army. If the unsuccessful battle and flight of the Gauls disquieted any, these, if they made inquiries, might discover that, when the Gauls had been tired out by the long duration of the war, Ariovistus, after he had many months kept himself in his camp and in the marshes, and had given no opportunity for an engagement, fell suddenly upon them, by this time despairing of a battle and scattered in all directions, and was victorious more through stratagem and cunning than valour. But though there had been room for such stratagem against savage and unskilled men, not even [Ariovistus] himself expected that thereby our armies could be entrapped. That those who ascribed their fear to a pretence about the [deficiency of] supplies and the narrowness of the roads, acted presumptuously, as they seemed either to distrust their general's discharge of his duty, or to dictate to him. That these things were his concern; that the Sequani, the Leuci, and the Lingones were to furnish the corn; and that it was already ripe in the fields; that as to the road they would soon be able to judge for themselves. As to its being reported that the soldiers would not be obedient to command, or advance, he was not at all disturbed at that; for he knew that in the case of all those whose army had not been obedient to command, either upon some mismanagement of an affair, fortune had deserted them, or, that upon some crime being discovered, covetousness had been clearly proved [against them]. His integrity had been seen throughout his whole life, his good fortune in the war with the Helvetii. That he would therefore instantly set about what he had intended to put off till a more distant day, and would break up his camp the next night, in the fourth watch, that he might ascertain, as soon as possible, whether a sense of honour and duty, or whether fear had more influence with them. But that, if no one else should follow, yet he would go with only the tenth legion, of which he had no misgivings, and it should be his praetorian cohort." — This legion Caesar had both greatly favoured, and in it, on account of its valour, placed the greatest confidence.

XLI. — Upon the delivery of this speech, the minds of all were changed in a surprising manner, and the highest ardour and eagerness for prosecuting the war were engendered; and the tenth legion was the first to return thanks to him, through their military tribunes, for his having expressed this most favourable opinion of them; and assured him that they were quite ready to prosecute the war. Then, the other legions endeavoured, through their military tribunes and the centurions of the principal companies, to excuse themselves to Caesar, [saying] that they had never either doubted or feared, or supposed that the determination of the conduct of the war was theirs and not their general's. Having accepted their excuse, and having had the road carefully reconnoitred by Divitiacus, because in him of all others he had the greatest faith, [he found] that by a circuitous route of more than fifty miles he might lead his army through open parts; he then set out in the fourth watch, as he had said [he would]. On the seventh day, as he did not discontinue his march, he was informed by scouts that the forces of Ariovistus were only four and twenty miles distant from ours.

XLII. — Upon being apprised of Caesar's arrival, Ariovistus sends ambassadors to him, [saying] that what he had before requested as to a conference, might now, as far as his permission went, take place, since he [Caesar] had approached nearer, and he considered that he might now do it without danger. Caesar did not reject the proposal and began to think that he was now returning to a rational state of mind, as he spontaneously proffered that which he had previously refused to him when requesting it; and was in great hopes that, in consideration of his own and the Roman people's great favours towards him, the issue would be that he would desist from his obstinacy upon his demands being made known. The fifth day after that was appointed as the day of conference. Meanwhile, as ambassadors were being often sent to and fro between them, Ariovistus demanded that Caesar should not bring any foot-soldier with him to the conference, [saying] that "he was afraid of being ensnared by him through treachery; that both should come accompanied by cavalry; that he would not come on any other condition." Caesar, as he neither wished that the conference should, by an excuse thrown in the way, be set aside, nor durst trust his life to the cavalry of the Gauls, decided that it would be most expedient to take away from the Gallic cavalry all their horses, and thereon to mount the legionary soldiers of the tenth legion, in which he placed the greatest confidence; in order that he might have a body-guard as trustworthy as possible, should there be any need for action. And when this was done, one of the soldiers of the tenth legion said, not without a touch of humour, "that Caesar did more for them than he had promised; he had promised to have the tenth legion in place of his praetorian cohort; but he now converted them into horse."

XLIII. — There was a large plain, and in it a mound of earth of considerable size. This spot was at nearly an equal distance from both camps. Thither, as had been appointed, they came for the conference. Caesar stationed the legion, which he had brought [with him] on horseback, 200 paces from this mound. The cavalry of Ariovistus also took their stand at an equal distance. Ariovistus then demanded that they should confer on horseback, and that, besides themselves, they should bring with them ten men each to the conference. When they were come to the place, Caesar, in the opening of his speech, detailed his own and the senate's favours towards him [Ariovistus], "in that he had been styled king, in that [he had been styled] friend, by the senate — in that very considerable presents had been sent him; which circumstance he informed him had both fallen to the lot of few, and had usually been bestowed in consideration of important personal services; that he, although he had neither an introduction, nor a just ground for the request, had obtained these honours through the kindness and munificence of himself [Caesar] and the senate. He informed him too, how old and how just were the grounds of connexion that existed between themselves [the Romans] and the Aedui, what decrees of the senate had been passed in their favour, and how frequent and how honourable; how from time immemorial the Aedui had held the supremacy of the whole of Gaul; even [said Caesar] before they had sought our friendship; that it was the custom of the Roman people to desire not only that its allies and friends should lose none of their property, but be advanced in influence, dignity, and honour: who then could endure that what they had brought with them to the friendship of the Roman people, should be torn from them?" He then made the same demands which he had commissioned the ambassadors to make, that [Ariovistus] should not make war either upon the Aedui or their allies, that he should restore the hostages; that, if he could not send back to their country any part of the Germans, he should at all events suffer none of them any more to cross the Rhine.

XLIV. — Ariovistus replied briefly to the demands of Caesar; but expatiated largely on his own virtues, "that he had crossed the Rhine not of his own accord, but on being invited and sent for by the Gauls; that he had not left home and kindred without great expectations and great rewards; that he had settlements in Gaul, granted by the Gauls themselves; that the hostages had been given by their own good-will; that he took by right of war the tribute which conquerors are accustomed to impose on the conquered; that he had not made war upon the Gauls, but the Gauls upon him; that all the states of Gaul came to attack him, and had encamped against him; that all their forces had been routed and beaten by him in a single battle; that if they chose to make a second trial, he was ready to encounter them again; but if they chose to enjoy peace, it was unfair to refuse the tribute, which of their own free-will they had paid up to that time. That the friendship of the Roman people ought to prove to him an ornament and a safeguard, not a detriment; and that he sought it with that expectation. But if through the Roman people the tribute was to be discontinued, and those who surrendered to be seduced from him, he would renounce the friendship of the Roman people no less heartily than he had sought it. As to his leading over a host of Germans into Gaul, that he was doing this with a view of securing himself, not of assaulting Gaul: that there was evidence of this, in that he did not come without being invited, and in that he did not make war, but merely warded it off. That he had come into Gaul before the Roman people. That never before this time did a Roman army go beyond the frontiers of the province of Gaul. What [said he] does [Caesar] desire? — why come into his [Ariovistus's] domains? — that this was his province of Gaul, just as that is ours. As it ought not to be pardoned in him, if he were to make an attack upon our territories; so, likewise, that we were unjust to obstruct him in his prerogative. As for Caesar's saying that the Aedui had been styled 'brethren' by the senate, he was not so uncivilized nor so ignorant of affairs, as not to know that the Aedui in the very last war with the Allobroges had neither rendered assistance to the Romans, nor received any from the Roman people in the struggles which the Aedui had been maintaining with him and with the Sequani. He must feel suspicious that Caesar, though feigning friendship as the reason for his keeping an army in Gaul; was keeping it with the view of crushing him. And that unless he depart, and withdraw his army from these parts, he shall regard him not as a friend, but as a foe; and that, even if he should put him to death, he should do what would please many of the nobles and leading men of the Roman people; he had assurance of that from themselves through their messengers, and could purchase the favour and the friendship of them all by his [Caesar's] death. But if he would depart and resign to him the free possession of Gaul, he would recompense him with a great reward, and would bring to a close whatever wars he wished to be carried on, without any trouble or risk to him."

XLV. — Many things were stated by Caesar to the effect [to show]: "why he could not waive the business, and that neither his nor the Roman people's practice would suffer him to abandon most meritorious allies, nor did he deem that Gaul belonged to Ariovistus rather than to the Roman people; that the Arverni and the Ruteni had been subdued in war by Quintus Fabius Maximus, and that the Roman people had pardoned them and had not reduced them into a province or imposed a tribute upon them. And if the most ancient period was to be regarded — then was the sovereignty of the Roman people in Gaul most just: if the decree of the senate was to be observed, then ought Gaul to be free, which they [the Romans] had conquered in war, and had permitted to enjoy its own laws."

XLVI. — While these things are being transacted in the conference, it was announced to Caesar that the cavalry of Ariovistus were approaching nearer the mound, and were riding up to our men, and casting stones and weapons at them. Caesar made an end of his speech and betook himself to his men; and commanded them that they should by no means return a weapon upon the enemy. For though he saw that an engagement with the cavalry would be without any danger to his chosen legion, yet he did not think proper to engage, lest, after the enemy were routed, it might be said that they had been ensnared by him under the sanction of a conference. When it was spread abroad among the common soldiery with what haughtiness Ariovistus had behaved at the conference, and how he had ordered the Romans to quit Gaul, and how his cavalry had made an attack upon our men, and how this had broken off the conference, a much greater alacrity and eagerness for battle was infused into our army.

XLVII. — Two days after, Ariovistus sends ambassadors to Caesar, to state "that he wished to treat with him about those things which had been begun to be treated of between them, but had not been concluded"; [and to beg] that "he would either again appoint a day for a conference; or, if he were not willing to do that, that he would send one of his [officers] as an ambassador to him." There did not appear to Caesar any good reason for holding a conference; and the more so as the day before the Germans could not be restrained from casting weapons at our men. He thought he should not without great danger send to him as ambassador one of his [Roman] officers, and should expose him to savage men. It seemed [therefore] most proper to send to him C. Valerius Procillus, the son of C. Valerius Caburus, a young man of the highest courage and accomplishments (whose father had been presented with the freedom of the city by C. Valerius Flaccus), both on account of his fidelity and on account of his knowledge of the Gallic language, which Ariovistus, by long practice, now spoke fluently; and because in his case the Germans would have no motive for committing violence; and [as his colleague] M. Mettius, who had shared the hospitality of Ariovistus. He commissioned them to learn what Ariovistus had to say, and to report to him. But when Ariovistus saw them before him in his camp, he cried out in the presence of his army, "Why were they come to him? was it for the purpose of acting as spies?" He stopped them when attempting to speak, and cast them into chains.

XLVIII. — The same day he moved his camp forward and pitched under a hill six miles from Caesar's camp. The day following he led his forces past Caesar's camp, and encamped two miles beyond him; with this design — that he might cut off Caesar from, the corn and provisions which might be conveyed to him from the Sequani and the Aedui. For five successive days from that day, Caesar drew out his forces before the camp, and put them in battle order, that, if Ariovistus should be willing to engage in battle, an opportunity might not be wanting to him. Ariovistus all this time kept his army in camp: but engaged daily in cavalry skirmishes. The method of battle in which the Germans had practised themselves was this: There were 6000 horse, and as many very active and courageous foot, one of whom each of the horse selected out of the whole army for his own protection. By these [foot] they were constantly accompanied in their engagements; to these the horse retired; these on any emergency rushed forward; if any one, upon receiving a very severe wound, had fallen from his horse, they stood around him: if it was necessary to advance farther: than usual, or to retreat more rapidly, so great, from practice, was their swiftness, that, supported by the manes of the horses, they could keep pace with their speed.

XLIX. — Perceiving that Ariovistus kept himself in camp, Caesar, that he might not any longer be cut off from provisions, chose a convenient position for a camp beyond that place in which the Germans had encamped, at about 600 paces from them, and having drawn up his army in three lines, marched to that place. He ordered the first and second lines to be under arms; the third to fortify the camp. This place was distant from the enemy about 600 paces, as has been stated. Thither Ariovistus sent light troops, about 16,000 men in number, with all his cavalry; which forces were to intimidate our men, and hinder them in their fortification. Caesar nevertheless, as he had before arranged, ordered two lines to drive off the enemy: the third to execute the work. The camp being fortified, he left there two legions and a portion of the auxiliaries; and led back the other four legions into the larger camp.

L. — The next day, according to his custom, Caesar led out his forces from both camps, and having advanced a little from the larger one, drew up his line of battle, and gave the enemy an opportunity of fighting. When he found that they did not even then come out [from their entrenchments], he led back his army into camp about noon. Then at last Ariovistus sent part of his forces to attack the lesser camp. The battle was vigorously maintained on both sides till the evening. At sunset, after many wounds had been inflicted and received, Ariovistus led back his forces into camp. When Caesar inquired of his prisoners, wherefore Ariovistus did not come to an engagement, he discovered this to be the reason — that among the Germans it was the custom for their matrons to pronounce from lots and divination whether it were expedient that the battle should be engaged in or not; that they had said, "that it was not the will of heaven that the Germans should conquer, if they engaged in battle before the new moon."

LI. — The day following, Caesar left what seemed sufficient as a guard for both camps; [and then] drew up all the auxiliaries in sight of the enemy, before the lesser camp, because he was not very powerful in the number of legionary soldiers, considering the number of the enemy; that [thereby] he might make use of his auxiliaries for appearance. He himself, having drawn up his army in three lines, advanced to the camp of the enemy. Then at last of necessity the Germans drew their forces out of camp, and disposed them canton by canton, at equal distances, the Harudes, Marcomanni, Tribocci, Vangiones, Nemetes, Sedusii, Suevi; and surrounded their whole army with their chariots and waggons, that no hope might be left in flight. On these they placed their women, who, with dishevelled hair and in tears, entreated the soldiers, as they went forward to battle, not to deliver them into slavery to the Romans.

LII. — Caesar appointed over each legion a lieutenant and a questor, that every one might have them as witnesses of his valour. He himself began the battle at the head of the right wing, because he had observed that part of the enemy to be the least strong. Accordingly our men, upon the signal being given, vigorously made an attack upon the enemy, and the enemy so suddenly and rapidly rushed forward, that there was no time for casting the javelins at them. Throwing aside [therefore] their javelins, they fought with swords hand to hand. But the Germans, according to their custom, rapidly forming a phalanx, sustained the attack of our swords. There were found very many of our soldiers who leaped upon the phalanx, and with their hands tore away the shields, and wounded the enemy from above. Although the army of the enemy was routed on the left wing and put to flight, they [still] pressed heavily on our men from the right wing, by the great number of their troops. On observing which, P. Crassus, a young man, who commanded the cavalry — as he was more disengaged than those who were employed in the fight — sent the third line as a relief to our men who were in distress.

LIII. — Thereupon the engagement was renewed, and all the enemy turned their backs, nor did they cease to flee until they arrived at the river Rhine, about fifty miles from that place. There some few, either relying on their strength, endeavoured to swim over, or, finding boats, procured their safety. Among the latter was Ariovistus, who meeting with a small vessel tied to the bank, escaped in it: our horse pursued and slew all the rest of them. Ariovistus had two wives, one a Suevan by nation, whom he had brought with him from home; the other a Norican, the sister of king Vocion, whom he had married in Gaul, she having been sent [thither for that purpose] by her brother. Both perished in that flight. Of their two daughters, one was slain, the other captured. C. Valerius Procillus, as he was being dragged by his guards in the flight, bound with a triple chain, fell into the hands of Caesar himself, as he was pursuing the enemy with his cavalry. This circumstance indeed afforded Caesar no less pleasure than the victory itself; because he saw a man of the first rank in the province of Gaul, his intimate acquaintance and friend, rescued from the hand of the enemy, and restored to him, and that fortune had not diminished aught of the joy and exultation [of that day] by his destruction. He [Procillus] said that in his own presence the lots had been thrice consulted respecting him, whether he should immediately be put to death by fire, or be reserved for another time: that by the favour of the lots he was uninjured. M. Mettius, also, was found and brought back to him [Caesar].

LIV.—This battle having been reported beyond the Rhine, the Suevi, who had come to the banks of that river, began to return home, when the Ubii, who dwelt nearest to the Rhine, pursuing them, while much alarmed, slew a great number of them. Caesar having concluded two very important wars in one campaign, conducted his army into winter-quarters among the Sequani, a little earlier than the season of the year required. He appointed Labienus over the winter-quarters, and set out in person for Hither Gaul to hold the assizes.

Book II

I. — While Caesar was in winter-quarters in Hither Gaul, as we have shown above, frequent reports were brought to him, and he was also informed by letters from Labienus, that all the Belgae, who we have said are a third part of Gaul, were entering into a confederacy against the Roman people, and giving hostages to one another; that the reasons of the confederacy were these — first, because they feared that, after all [Celtic] Gaul was subdued, our army would be led against them; secondly, because they were instigated by several of the Gauls; some of whom as [on the one hand] they had been unwilling that the Germans should remain any longer in Gaul, so [on the other] they were dissatisfied that the army of the Roman people should pass the winter in it, and settle there; and others of them, from a natural instability and fickleness of disposition, were anxious for a revolution; [the Belgae were instigated] by several, also, because the government in Gaul was generally seized upon by the more powerful persons and by those who had the means of hiring troops, and they could less easily effect this object under our dominion.

II. — Alarmed by these tidings and letters, Caesar levied two new legions in Hither Gaul, and, at the beginning of summer, sent Q. Pedius, his lieutenant, to conduct them further into Gaul. He himself, as soon as there began to be plenty of forage, came to the army. He gives a commission to the Senones and the other Gauls who were neighbours of the Belgae, to learn what is going on amongst them [*i.e.* the Belgae], and inform him of these matters. These all uniformly reported that troops were being raised, and that an army was being collected in one place. Then, indeed, he thought that he ought not to hesitate about proceeding towards them, and having provided supplies, moves his camp, and in about fifteen days arrives at the territories of the Belgae.

III. — As he arrived there unexpectedly and sooner than any one anticipated, the Remi, who are the nearest of the Belgae to [Celtic] Gaul, sent to him Iccius and Antebrogius, [two of] the principal persons of the state, as their ambassadors: to tell him that they surrendered themselves and all their possessions to the protection and disposal of the Roman people: and that they had neither combined with the rest of the Belgae, nor entered into any confederacy against the Roman people: and were prepared to give hostages, to obey his commands, to receive him into their towns, and to aid him with corn and other things; that all the rest of the Belgae were in arms; and that the Germans, who dwell on this side the Rhine, had joined themselves to them; and that so great was the infatuation of them all that they could not restrain even the Suessiones, their own brethren and kinsmen, who enjoy the same rights, and the same laws, and who have one government and one magistracy [in common] with themselves, from uniting with them.

IV. — When Caesar inquired of them what states were in arms, how powerful they were, and what they could do in war, he received the following information: that the greater part of the Belgae were sprung from the Germans, and that having crossed the Rhine at an early period, they had settled there, on account of the fertility of the country, and had driven out the Gauls who inhabited those regions; and that they were the only people who, in the memory of our fathers, when all Gaul was overrun, had prevented the Teutones and the Cimbri from entering their territories; the effect of which was that, from the recollection of those events, they assumed to themselves great authority and haughtiness in military matters. The Remi said that they had known accurately everything respecting their number, because, being united to them by neighbourhood and by alliances, they had learnt what number each state had in the general council of the Belgae promised for that war. That the Bellovaci were the most powerful amongst them in valour, influence, and number of men; that these could muster 100,000 armed men, [and had] promised 60,000 picked men out of that number, and demanded for themselves the command of the whole war. That the Suessiones were their nearest neighbours and possessed a very extensive and fertile country; that among them, even in our own memory, Divitiacus, the most powerful man of all Gaul, had been king; who had held the government of a great part of these regions, as well as of Britain; that their king at present was Galba; that the direction of the whole war was conferred by the consent of all upon him, on account of his integrity and prudence; that they had twelve towns; that they had promised 50,000 armed men; and that the Nervii, who are reckoned the most warlike among them, and are situated at a very great distance, [had promised] as many; the Atrebates, 15,000; the Ambiani, 10,000; the Morini, 25,000; the Menapu, 9,000; the Caleti, 10,000; the Velocasses and the Veromandui as many; the Aduatuci, 19,000; that the Condrusi, the Eburones, the Caeraesi, the Paemani, who are called by the common name of Germans, [had promised], they thought, to the number of 40,000.

V. — Caesar, having encouraged the Remi, and addressed them courteously, ordered the whole senate to assemble before him, and the children of their chief men to be brought to him as hostages; all which commands they punctually performed by the day [appointed]. He, addressing himself to Divitiacus the Aeduan, with great earnestness, points out how much it concerns the republic and their common security, that the forces of the enemy should be divided, so that it might not be necessary to engage with so large a number at one time. [He asserts] that this might be effected if the Aedui would lead their forces into the territories of the Bellovaci, and begin to lay waste their country. With these instructions he dismissed him from his presence. After he perceived that all the forces of the Belgae, which had been collected in one place, were approaching towards him, and learnt from the scouts whom he had sent out, and [also] from the Remi, that they were not then far distant, he hastened to lead his army over the Aisne, which is on the borders of the Remi, and there pitched his camp. This position fortified one side of his camp by the banks of the river, rendered the country which lay in his rear secure from the enemy, and furthermore ensured that provisions might without danger be brought to him by the Remi and the rest of the states. Over that river was a bridge: there he places a guard; and on the other side of the river he leaves Q. Titurus Sabinus, his lieutenant, with six cohorts. He orders him to fortify a camp with a rampart twelve feet in height, and a trench eighteen feet in breadth.

VI. — There was a town of the Remi, by name Bibrax, eight miles distant from this camp. This the Belgae on their march began to attack with great vigour. [The assault] was with difficulty sustained for that day. The Gauls' mode of besieging is the same as that of the Belgae: when after having drawn a large number of men around the whole of the fortifications, stones have begun to be cast against the wall on all sides, and the wall has been stript of its defenders, [then], forming a testudo, they advance to the gates and undermine the wall: which was easily effected on this occasion; for while so large a number were casting stones and darts, no one was able to maintain his position upon the wall. When night had put an end to the assault, Iccius, who was then in command of the town, one of the Remi, a man of the highest rank and influence amongst his people, and one of those who had come to Caesar as ambassador [to sue] for a peace, sends messengers to him, [to report] "That, unless assistance were sent to him, he could not hold out any longer."

VII. — Thither immediately after midnight, Caesar, using as guides the same persons who had come to him as messengers from Iccius, sends some Numidian and Cretan archers, and some Balearian slingers as a relief to the townspeople, by whose arrival both a desire to resist together with the hope of [making good their] defence was infused into the Remi, and, for the same reason, the hope of gaining the town abandoned the enemy. Therefore, after staying a short time before the town, and laying waste the country of the Remi, when all the villages and buildings which they could approach had been burnt, they hastened with all their forces to the camp of Caesar, and encamped within less than two miles [of it]; and their camp, as was indicated by the smoke and fires, extended more than eight miles in breadth.

VIII. — Caesar at first determined to decline a battle, as well on account of the great number of the enemy as their distinguished reputation for valour: daily, however, in cavalry actions, he strove to ascertain by frequent trials what the enemy could effect by their prowess and what our men would dare. When he perceived that our men were not inferior, as the place before the camp was naturally convenient and suitable for marshalling an army (since the hill where the camp was pitched, rising gradually from the plain, extended forward in breadth as far as the space which the marshalled army could occupy, and had steep declines of its side in either direction, and gently sloping in front gradually sank to the plain), on either side of that hill he drew a cross trench of about four hundred paces, and at the extremities of that trench built forts, and placed there his military engines, lest, after he had marshalled his army, the enemy, since they were so powerful in point of number, should be able to surround his men in the flank, while fighting. After doing this, and leaving in the camp the two legions which he had last raised, that, if there should be any occasion, they might be brought as a reserve, he formed the other six legions in order of battle before the camp. The enemy, likewise, had drawn up their forces which they had brought out of the camp.

IX. — There was a marsh of no great extent between our army and that of the enemy. The latter were waiting to see if our men would pass this; our men, also, were ready in arms to attack them while disordered, if the first attempt to pass should be made by them. In the meantime battle was commenced between the two armies by a cavalry action. When neither army began to pass the marsh, Caesar, upon the skirmishes of the horse [proving] favourable to our men, led back his forces into the camp. The enemy immediately hastened from that place to the river Aisne, which it has been stated was behind our camp. Finding a ford there, they endeavoured to lead a part of their forces over it; with the design, that, if they could, they might carry by storm the fort which Q. Titurius, Caesar's lieutenant, commanded, and might cut off the bridge; but, if they could not do that, they should lay waste the lands of the Remi, which were of great use to us in carrying on the war, and might hinder our men from foraging.

X. — Caesar, being apprised of this by Titurius, leads all his cavalry and light-armed Numidians, slingers and archers, over the bridge, and hastens towards them. There was a severe struggle in that place. Our men, attacking in the river the disordered enemy, slew a great part of them. By the immense number of their missiles they drove back the rest, who in a most courageous manner were attempting to pass over their bodies, and surrounded with their cavalry, and cut to pieces those who had first crossed the river. The enemy, when they perceived that their hopes had deceived them both with regard to their taking the town by storm and also their passing the river, and did not see our men advance to a more disadvantageous place for the purpose of fighting, and when provisions began to fail them, having called a council, determined that it was best for each to return to his country, and resolved to assemble from all quarters to defend those into whose territories the Romans should first march an army; that they might contend in their own rather than in a foreign country, and might enjoy the stores of provisions which they possessed at home. Together with other causes, this consideration also led them to that resolution, viz.: that they had learnt that Divitiacus and the Aedui were approaching the territories of the Bellovaci. And it was impossible to persuade the latter to stay any longer, or to deter them from conveying succour to their own people.

XI. — That matter being determined on, marching out of their camp at the second watch, with great noise and confusion, in no fixed order, nor under any command, since each sought for himself the foremost place in the journey, and hastened to reach home, they made their departure appear very like a flight. Caesar, immediately learning this through his scouts, [but] fearing an ambuscade, because he had not yet discovered for what reason they were departing, kept his army and cavalry within the camp. At daybreak, the intelligence having been confirmed by the scouts, he sent forward his cavalry to harass their rear; and gave the command of it to two of his lieutenants, Q. Pedius, and L. Aurunculeius Cotta. He ordered T. Labienus, another of his lieutenants, to follow them closely with three legions. These, attacking their rear, and pursuing them for many miles, slew a great number of them as they were fleeing; while those in the rear with whom they had come up, halted, and bravely sustained the attack of our soldiers; the van, because they appeared to be removed from danger, and were not restrained by any necessity or command, as soon as the noise was heard, broke their ranks, and, to a man, rested their safety in flight. Thus without any risk [to themselves] our men killed as great a number of them as the length of the day allowed; and at sunset desisted from the pursuit, and betook themselves into the camp, as they had been commanded.

XII. — On the day following, before the enemy could recover from their terror and flight, Caesar led his army into the territories of the Suessiones, which are next to the Remi, and having accomplished a long march, hastens to the town named Noviodunum. Having attempted to take it by storm on his march, because he heard that it was destitute of [sufficient] defenders, he was not able to carry it by assault, on account of the breadth of the ditch and the height of the wall, though few were defending it. Therefore, having fortified the camp, he began to bring up the vineae, and to provide whatever things were necessary for the storm. In the meantime, the whole body of the Suessiones, after their flight, came the next night into the town. The vineae having been quickly brought up against the town, a mound thrown up, and towers built, the Gauls, amazed by the greatness of the works, such as they had neither seen nor heard of before, and struck, also, by the despatch of the Romans, send ambassadors to Caesar respecting a surrender, and succeed in consequence of the Remi requesting that they [the Suessiones] might be spared.

XIII. — Caesar, having received as hostages the first men of the state, and even the two sons of king Galba himself; and all the arms in the town having been delivered up, admitted the Suessiones to a surrender, and led his army against the Bellovaci. Who, when they had conveyed themselves and all their possessions into the town called Bratuspantium, and Caesar with his army was about five miles distant from that town, all the old men, going out of the town, began to stretch out their hands to Caesar, and to intimate by their voice that they would throw themselves on his protection and power, nor would contend in arms against the Roman people. In like manner, when he had come up to the town, and there pitched his camp, the boys and the women from the wall, with outstretched hands, after their custom, begged peace from the Romans.

XIV. — For these Divitiacus pleads (for after the departure of the Belgae, having dismissed the troops of the Aedui, he had returned to Caesar). "The Bellovaci had at all times been in the alliance and friendship of the Aeduan state; that they had revolted from the Aedui and made war upon the Roman people, being urged thereto by their nobles, who said that the Aedui, reduced to slavery by Caesar, were suffering every indignity and insult. That they who had been the leaders of that plot, because they perceived how great a calamity they had brought upon the state, had fled into Britain. That not only the Bellovaci, but also the Aedui, entreated him to use his [accustomed] clemency and lenity towards them [the Bellovaci]: which if he did, he would increase the influence of the Aedui among all the Belgae, by whose succour and resources they had been accustomed to support themselves whenever any wars occurred."

XV. — Caesar said that on account of his respect for Divitiacus and the Aeduans, he would receive them into his protection, and would spare them; but, because the state was of great influence among the Belgae, and pre-eminent in the number of its population, he demanded 600 hostages. When these were delivered, and all the arms in the town collected, he went from that place into the territories of the Ambiani, who, without delay, surrendered themselves and all their possessions. Upon their territories bordered the Nervii, concerning whose character and customs when Caesar inquired he received the following information: — That "there was no access for merchants to them; that they suffered no wine and other things tending to luxury to be imported; because they thought that by their use the mind is enervated and the courage impaired: that they were a savage people and of great bravery: that they upbraided and condemned the rest of the Belgae who had surrendered themselves to the Roman people and thrown aside their national courage: that they openly declared they would neither send ambassadors, nor accept any condition of peace."

XVI. — After he had made three days' march through their territories, he discovered from some prisoners, that the river Sambre was not more than ten miles from his camp: that all the Nervii had stationed themselves on the other side of that river, and together with the Atrebates and the Veromandui, their neighbours, were there awaiting the arrival of the Romans; for they had persuaded both these nations to try the same fortune of war [as themselves]: that the forces of the Aduatuci were also expected by them, and were on their march; that they had put their women, and those who through age appeared useless for war, in a place to which there was no approach for an army, on account of the marshes.

XVII. — Having learnt these things, he sends forward scouts and centurions to choose a convenient place for the camp. And as a great many of the surrounding Belgae and other Gauls, following Caesar, marched with him; some of these, as was afterwards learnt from the prisoners, having accurately observed, during those days, the army's method of marching, went by night to the Nervii, and informed them that a great number of baggage-trains passed between the several legions, and that there would be no difficulty, when the first legion had come into the camp, and the other legions were at a great distance, to attack that legion while under baggage, which being routed, and the baggage-train seized, it would come to pass that the other legions would not dare to stand their ground. It added weight also to the advice of those who reported that circumstance, that the Nervii, from early times, because they were weak in cavalry (for not even at this time do they attend to it, but accomplish by their infantry whatever they can), in order that they might the more easily obstruct the cavalry of their neighbours if they came upon them for the purpose of plundering, having cut young trees, and bent them, by means of their numerous branches [extending] on to the sides, and the quick-briars and thorns springing up between them, had made these hedges present a fortification like a wall, through which it was not only impossible to enter, but even to penetrate with the eye. Since [therefore] the march of our army would be obstructed by these things, the Nervii thought that the advice ought not to be neglected by them.

XVIII. — The nature of the ground which our men had chosen for the camp was this: A hill, declining evenly from the top, extended to the river Sambre, which we have mentioned above: from this river there arose a [second] hill of like ascent, on the other side and opposite to the former, and open from about 200 paces at the lowest part; but in the upper part, woody, (so much so) that it was not easy to see through it into the interior. Within those woods the enemy kept themselves in concealment; a few troops of horse-soldiers appeared on the open ground, along the river. The depth of the river was about three feet.

XIX. — Caesar, having sent his cavalry on before, followed close after them with all his forces; but the plan and order of the march was different from that which the Belgae had reported to the Nervii. For as he was approaching the enemy Caesar, according to his custom, led on [as the van] six legions unencumbered by baggage; behind them he had placed the baggage-trains of the whole army; then the two legions which had been last raised closed the rear, and were a guard for the baggage-train. Our horse, with the slingers and archers, having passed the river, commenced action with the cavalry of the enemy. While they from time to time betook themselves into the woods to their companions, and again made an assault out of the wood upon our men, who did not dare to follow them in their retreat further than the limit to which the plain and open parts extended, in the meantime the six legions which had arrived first, having measured out the work, began to fortify the camp. When the first part of the baggage-train of our army was seen by those who lay hid in the woods, which had been agreed on among them as the time for commencing action, as soon as they had arranged their line of battle and formed their ranks within the woods, and had encouraged one another, they rushed out suddenly with all their forces and made an attack upon our horse. The latter being easily routed and thrown into confusion, the Nervii ran down to the river with such incredible speed that they seemed to be in the woods, the river, and close upon us almost at the same time. And with the same speed they hastened up the hill to our camp and to those who were employed in the works.

XX. — Caesar had everything to do at one time: the standard to be displayed, which was the sign when it was necessary to run to arms; the signal to be given by the trumpet; the soldiers to be called off from the works; those who had proceeded some distance for the purpose of seeking materials for the rampart, to be summoned; the order of battle to be formed; the soldiers to be encouraged; the watchword to be given. A great part of these arrangements was prevented by the shortness of time and the sudden approach and charge of the enemy. Under these difficulties two things proved of advantage; [first] the skill and experience of the soldiers, because, having been trained by former engagements, they could suggest to themselves what ought to be done, as conveniently as receive information from others; and [secondly] that Caesar had forbidden his several lieutenants to depart from the works and their respective legions, before the camp was fortified. These, on account of the near approach and the speed of the enemy, did not then wait for any command from Caesar, but of themselves executed whatever appeared proper.

XXI. — Caesar, having given the necessary orders, hastened to and fro into whatever quarter fortune carried him to animate the troops, and came to the tenth legion. Having encouraged the soldiers with no further speech than that "they should keep up the remembrance of their wonted valour, and not be confused in mind, but valiantly sustain the assault of the enemy"; as the latter were not farther from them than the distance to which a dart could be cast, he gave the signal for commencing battle. And having gone to another quarter for the purpose of encouraging [the soldiers], he finds them fighting. Such was the shortness of the time, and so determined was the mind of the enemy on fighting, that time was wanting not only for affixing the military insignia, but even for putting on the helmets and drawing off the covers from the shields. To whatever part any one by chance came from the works (in which he had been employed), and whatever standards he saw first, at these he stood, lest in seeking his own company he should lose the time for fighting.

XXII. — The army having been marshalled, rather as the nature of the ground and the declivity of the hill and the exigency of the time, than as the method and order of military matters required; whilst the legions in the different places were withstanding the enemy, some in one quarter, some in another, and the view was obstructed by the very thick hedges intervening, as we have before remarked, neither could proper reserves be posted, nor could the necessary measures be taken in each part, nor could all the commands be issued by one person. Therefore, in such an unfavourable state of affairs, various events of fortune followed.

XXIII. — The soldiers of the ninth and tenth legions, as they had been stationed on the left part of the army, casting their weapons, speedily drove the Atrebates (for that division had been opposed to them), who were breathless with running and fatigue, and worn out with wounds, from the higher ground into the river; and following them as they were endeavouring to pass it, slew with their swords a great part of them while impeded (therein). They themselves did not hesitate to pass the river; and having advanced to a disadvantageous place, when the battle was renewed, they [nevertheless] again put to flight the enemy, who had returned and were opposing them. In like manner, in another quarter two different legions, the eleventh and the eighth, having routed the Veromandui, with whom they had engaged, were fighting from the higher ground upon the very banks of the river. But, almost the whole camp on the front and on the left side being then exposed, since the twelfth legion was posted in the right wing, and the seventh at no great distance from it, all the Nervii, in a very close body, with Boduognatus, who held the chief command, as their leader, hastened towards that place; and part of them began to surround the legions on their unprotected flank, part to make for the highest point of the encampment.

XXIV. — At the same time our horsemen, and light-armed infantry, who had been with those who, as I have related, were routed by the first assault of the enemy, as they were betaking themselves into the camp, met the enemy face to face, and again sought flight into another quarter; and the camp-followers who from the Decuman Gate and from the highest ridge of the hill had seen our men pass the river as victors, when, after going out for the purposes of plundering, they looked back and saw the enemy parading in our camp, committed themselves precipitately to flight; at the same time there arose the cry and shout of those who came with the baggage-train; and they (affrighted) were carried some one way, some another. By all these circumstances the cavalry of the Treviri were much alarmed (whose reputation for courage is extraordinary among the Gauls, and who had come to Caesar, being sent by their state as auxiliaries), and, when they saw our camp filled with a large number of the enemy, the legions hard pressed and almost held surrounded, the camp-retainers, horsemen, slingers, and Numidians fleeing on all sides divided and scattered, they, despairing of our affairs, hastened home, and related to their state that the Romans were routed and conquered, [and] that the enemy were in possession of their camp and baggage-train.

XXV. — Caesar proceeded, after encouraging the tenth legion, to the right wing; where he perceived that his men were hard pressed, and that in consequence of the standards of the twelfth legion being collected together in one place, the crowded soldiers were a hindrance to themselves in the fight; that all the centurions of the fourth cohort were slain, and the standard-bearer killed, the standard itself lost, almost all the centurions of the other cohorts either wounded or slain, and among them the chief centurion of the legion, P. Sextius Baculus, a very valiant man, who was so exhausted by many and severe wounds, that he was already unable to support himself; he likewise perceived that the rest were slackening their efforts, and that some, deserted by those in the rear, were retiring from the battle and avoiding the weapons; that the enemy [on the other hand], though advancing from the lower ground, were not relaxing in front, and were [at the same time] pressing hard on both flanks; he also perceived that the affair was at a crisis, and that there was not any reserve which could be brought up; having therefore snatched a shield from one of the soldiers in the rear (for he himself had come without a shield), he advanced to the front of the line, and addressing the centurions by name, and encouraging the rest of the soldiers, he ordered them to carry forward the standards, and extend the companies, that they might the more easily use their swords. On his arrival, as hope was brought to the soldiers and their courage restored, whilst every one for his own part, in the sight of his general, desired to exert his utmost energy, the impetuosity of the enemy was a little checked.

XXVI. — Caesar, when he perceived that the seventh legion, which stood close by him, was also hard pressed by the enemy, directed the tribunes of the soldiers to effect a junction of the legions gradually, and make their charge upon the enemy with a double front; which having been done, since they brought assistance the one to the other, nor feared lest their rear should be surrounded by the enemy, they began to stand their ground more boldly, and to fight more courageously. In the meantime, the soldiers of the two legions which had been in the rear of the army, as a guard for the baggage-train, upon the battle being reported to them, quickened their pace, and were seen by the enemy on the top of the hill; and Titus Labienus, having gained possession of the camp of the enemy, and observed from the higher ground what was going on in our camp, sent the tenth legion as a relief to our men, who, when they had learnt from the flight of the horse and the sutlers in what position the affair was, and in how great danger the camp and the legion and the commander were involved, left undone nothing [which tended] to despatch.

XXVI. — By their arrival, so great a change of matters was made, that our men, even those who had fallen down exhausted with wounds, leant on their shields, and renewed the fight: then the camp-retainers, though unarmed, seeing the enemy completely dismayed, attacked [them though] armed; the horsemen too, that they might by their valour blot out the disgrace of their flight, thrust themselves before the legionary soldiers in all parts of the battle. But the enemy, even in the last hope of safety, displayed such great courage that when the foremost of them had fallen, the next stood upon them prostrate, and fought from their bodies; when these were overthrown, and their corpses heaped up together, those who survived cast their weapons against our men [thence], as from a mound, and returned our darts which had fallen between [the armies]; so that it ought not to be concluded, that men of such great courage had injudiciously dared to pass a very broad river, ascend very high banks, and come up to a very disadvantageous place; since their greatness of spirit had rendered these actions easy, although in themselves very difficult.

XXVIII. — This battle being ended, and the nation and name of the Nervii being almost reduced to annihilation, their old men, whom together with the boys and women we have stated to have been collected together in the fenny places and marshes, on this battle having been reported to them, since they were convinced that nothing was an obstacle to the conquerors, and nothing safe to the conquered, sent ambassadors to Caesar by the consent of all who remained, and surrendered themselves to him; and in recounting the calamity of their state, said that their senators were reduced from 600 to three; that from 60,000 men they [were reduced] to scarcely 500 who could bear arms; whom Caesar, that he might appear to use compassion towards the wretched and the suppliant, most carefully spared; and ordered them to enjoy their own territories and towns, and commanded their neighbours that they should restrain themselves and their dependants from offering injury or outrage [to them].

XXIX. — When the Aduatuci, of whom we have written above, were coming with all their forces to the assistance of the Nervii, upon this battle being reported to them, they returned home after they were on the march; deserting all their towns and forts, they conveyed together all their possessions into one town, eminently fortified by nature. While this town had on all sides around it very high rocks and precipices, there was left on one side a gently ascending approach, of not more than 200 feet in width; which place they had fortified with a very lofty double wall: besides, they had placed stones of great weight and sharpened stakes upon the walls. They were descended from the Cimbri and Teutones, who, when they were marching into our province and Italy, having deposited on this side the river Rhine such of their baggage-trains as they could not drive or convey with them, left 6,000 of their men as a guard and defence for them. These having, after the destruction of their countrymen, been harassed for many years by their neighbours, while one time they waged war offensively, and at another resisted it when waged against them, concluded a peace with the consent of all, and chose this place as their settlement.

XXX. — And on the first arrival of our army they made frequent sallies from the town, and contended with our men in trifling skirmishes: afterwards, when hemmed in by a rampart of twelve feet [in height], and fifteen miles in circuit, they kept themselves within the town. When, vineae having been brought up and a mound raised, they observed that a tower also was being built at a distance, they at first began to mock the Romans from their wall, and to taunt them with the following speeches. "For what purpose was so vast a machine constructed at so great a distance?" "With what hands," or "with what strength did they, especially [as they were] men of such very small stature" (for our shortness of stature, in comparison with the great size of their bodies, is generally a subject of much contempt to the men of Gaul), "trust to place against their walls a tower of such great weight."

XXXI. — But when they saw that it was being moved, and was approaching their walls, startled by the new and unaccustomed sight, they sent ambassadors to Caesar [to treat] about peace; who spoke in the following manner: "That they did not believe the Romans waged war without divine aid, since they were able to move forward machines of such a height with so great speed, and thus fight from close quarters: that they resigned themselves and all their possessions to [Caesar's] disposal: that they begged and earnestly entreated one thing, viz., that if perchance, agreeably to his clemency and humanity, which they had heard of from others, he should resolve that the Aduatuci were to be spared, he would not deprive them of their arms; that all their neighbours were enemies to them and envied their courage, from whom they could not defend themselves if their arms were delivered up: that it was better for them, if they should be reduced to that state, to suffer any fate from the Roman people, than to be tortured to death by those among whom they had been accustomed to rule."

XXXII. — To these things Caesar replied, "That he, in accordance with his custom, rather than owing to their desert, should spare the state, if they should surrender themselves before the battering-ram should touch the wall; but that there was no condition of surrender, except upon their arms being delivered up; that he should do to them that which he had done in the case of the Nervii, and would command their neighbours not to offer any injury to those who had surrendered to the Roman people." The matter being reported to their countrymen, they said that they would execute his commands. Having cast a very large quantity of their arms from the wall into the trench which was before the town, so that the heaps of arms almost equalled the top of the wall and the rampart, and nevertheless having retained and concealed, as we afterwards discovered, about a third part in the town, the gates were opened, and they enjoyed peace for that day.

XXXIII. — Towards evening Caesar ordered the gates to be shut, and the soldiers to go out of the town, lest the townspeople should receive any injury from them by night. They [the Aduatuci], by a design before entered into, as we afterwards understood, because they believed that, as a surrender had been made, our men would dismiss their guards, or at least would keep watch less carefully, partly with those arms which they had retained and concealed, partly with shields made of bark or interwoven wickers, which they had hastily covered over with skins (as the shortness of time required) in the third watch, suddenly made a sally from the town with all their forces [in that direction] in which the ascent to our fortifications seemed the least difficult. The signal having been immediately given by fires, as Caesar had previously commanded, a rush was made thither [i.e. by the Roman soldiers] from the nearest fort; and the battle was fought by the enemy as vigorously as it ought to be fought by brave men, in the last hope of safety, in a disadvantageous place, and against those who were throwing their weapons from a rampart and from towers; since all hope of safety depended on their courage alone. About 4,000 of the men having been slain, the rest were forced back into the town. The day after, Caesar, after breaking open the gates, which there was no one then to defend, and sending in our soldiers, sold the whole spoil of that town. The number of 53,000 persons was reported to him by those who had bought them.

XXXIV. — At the same time he was informed by P. Crassus, whom he had sent with one legion against the Veneti, the Unelli, the Osismii, the Curiosolitae, the Sesuvii, the Aulerci, and the Rhedones, which are maritime states, and touch upon the [Atlantic] ocean, that all these nations were brought under the dominion and power of the Roman people.

XXXV. — These things being achieved, [and] all Gaul being subdued, so high an opinion of this war was spread among the barbarians, that ambassadors were sent to Caesar by those nations who dwelt beyond the Rhine, to promise that they would give hostages and execute his commands. Which embassies Caesar, because he was hastening into Italy and Illyricum, ordered to return to him at the beginning of the following summer. He himself, having led his legions into winter-quarters among the Carnutes, the Andes, and the Turones, which states were close to those regions in which he had waged war, set out for Italy; and a thanksgiving of fifteen days was decreed for those achievements, upon receiving Caesar's letter; [an honour] which before that time had been conferred on none.

Book III

I. — When Caesar was setting out for Italy, he sent Servius Galba with the twelfth legion and part of the cavalry against the Nantuates, the Veragri, and Seduni, who extend from the territories of the Allobroges, and the lake of Geneva, and the river Rhone to the top of the Alps. The reason for sending him was, that he desired that the pass along the Alps, through which [the Roman] merchants had been accustomed to travel with great danger, and under great imposts, should be opened. He permitted him, if he thought it necessary, to station the legion in these places, for the purpose of wintering. Galba having fought some successful battles, and stormed several of their forts, upon ambassadors being sent to him from all parts and hostages given and a peace concluded, determined to station two cohorts among the Nantuates, and to winter in person with the other cohorts of that legion in a village of the Veragri, which is called Octodurus; and this village being situated in a valley, with a small plain annexed to it, is bounded on all sides by very high mountains. As this village was divided into two parts by a river, he granted one part of it to the Gauls, and assigned the other, which had been left by them unoccupied, to the cohorts to winter in. He fortified this [latter] part with a rampart and a ditch.

II. — When several days had elapsed in winter-quarters, and he had ordered corn to be brought in, he was suddenly informed by his scouts that all the people had gone off in the night from that part of the town which he had given up to the Gauls, and that the mountains which hung over it were occupied by a very large force of the Sedani and Veragri. It had happened for several reasons that the Gauls suddenly formed the design of renewing the war and cutting off that legion. First, because they despised a single legion, on account of its small number, and that not quite full (two cohorts having been detached, and several individuals being absent, who had been despatched for the purpose of seeking provision); then, likewise, because they thought that on account of the disadvantageous character of the situation, even their first attack could not be sustained [by us] when they would rush from the mountains into the valley, and discharge their weapons upon us. To this was added, that they were indignant that their children were torn from them under the title of hostages, and they were persuaded that the Romans designed to seize upon the summits of the Alps, and unite those parts to the neighbouring province [of Gaul], not only to secure the passes, but also as a constant possession.

III. — Having received these tidings, Galba, since the works of the winter-quarters and the fortifications were not fully completed, nor was sufficient preparation made with regard to corn and other provisions (since, as a surrender had been made, and hostages received, he had thought he need entertain no apprehension of a war), speedily summoning a council, began to anxiously inquire their opinions. In which council, since so much sudden danger had happened contrary to the general expectation, and almost all the higher places were seen already covered with a multitude of armed men, nor could [either] troops come to their relief, or provisions be brought in, as the passes were blocked up [by the enemy]; safety being now nearly despaired of, some opinions of this sort were delivered; that, "leaving their baggage, and making a sally, they should hasten away for safety by the same routes by which they had come thither." To the greater part, however, it seemed best, reserving that measure to the last, to await the issue of the matter, and to defend the camp.

IV. — A short time only having elapsed, so that time was scarcely given for arranging and executing those things which they had determined on, the enemy, upon the signal being given, rushed down [upon our men] from all parts, and discharged stones and darts upon our rampart. Our men at first, while their strength was fresh, resisted bravely, nor did they cast any weapon ineffectually from their higher station. As soon as any part of the camp, being destitute of defenders, seemed to be hard pressed, thither they ran, and brought assistance. But they were over-matched in this, that the enemy when wearied by the long continuance of the battle, went out of the action, and others with fresh strength came in their place; none of which things could be done by our men, owing to the smallness of their number; and not only was permission not given to the wearied [Roman] to retire from the fight, but not even to the wounded [was liberty granted] to quit the post where he had been stationed, and recover.

V. — When they had now been fighting for more than six hours, without cessation, and not only strength, but even weapons were failing our men, and the enemy were pressing on more rigorously, and had begun to demolish the rampart and to fill up the trench, while our men were becoming exhausted, and the matter was now brought to the last extremity, P. Sextius Baculus, a centurion of the first rank, whom we have related to have been disabled by severe wounds in the engagement with the Nervii, and also C. Volusenus, a tribune of the soldiers, a man of great skill and valour, hasten to Galba, and assure him that the only hope of safety lay in making a sally, and trying the last resource. Whereupon, assembling the centurions, he quickly gives orders to the soldiers to discontinue the fight a short time, and only collect the weapons flung [at them], and recruit themselves after their fatigue, and afterwards, upon the signal being given, sally forth from the camp, and place in their valour all their hope of safety.

VI. — They do what they were ordered; and, making a sudden sally from all the gates [of the camp], leave the enemy the means neither of knowing what was taking place, nor of collecting themselves. Fortune thus taking a turn, [our men] surround on every side, and slay those who had entertained the hope of gaining the camp, and having killed more than the third part of an army of more than 30,000 men (which number of the barbarians it appeared certain had come up to our camp), put to flight the rest when panic-stricken, and do not suffer them to halt even upon the higher grounds. All the forces of the enemy being thus routed, and stripped of their arms, [our men] betake themselves to their camp and fortifications. Which battle being finished, inasmuch as Galba was unwilling to tempt fortune again, and remembered that he had come into winter-quarters with one design, and saw that he had met with a different state of affairs; chiefly however urged by the want of corn and provision, having the next day burned all the buildings of that village, he hastens to return into the province; and as no enemy opposed or hindered his march, he brought the legion safe into the [country of the] Nantuates, thence into [that of] the Allobroges, and there wintered.

VII.—These things being achieved, while Caesar had every reason to suppose that Gaul was reduced to a state of tranquillity, the Belgae being overcome, the Germans expelled, the Seduni among the Alps defeated, and when he had, therefore, in the beginning of winter, set out for Illyricum, as he wished to visit those nations, and acquire a knowledge of their countries, a sudden war sprang up in Gaul. The occasion of that war was this: P. Crassus, a young man, had taken up his winter-quarters with the seventh legion among the Andes, who border upon the [Atlantic] ocean. He, as there was a scarcity of corn in those parts, sent out some officers of cavalry and several military tribunes amongst the neighbouring states, for the purpose of procuring corn and provision; in which number T. Terrasidius was sent amongst the Esubii; M. Trebius Gallus amongst the Curiosolitae; Q. Velanius, with T. Silius, amongst the Veneti.

VIII.—The influence of this state is by far the most considerable of any of the countries on the whole sea coast, because the Veneti both have a very great number of ships, with which they have been accustomed to sail to Britain, and [thus] excel the rest in their knowledge and experience of nautical affairs; and as only a few ports lie scattered along that stormy and open sea, of which they are in possession, they hold as tributaries almost all those who are accustomed to traffic in that sea. With them arose the beginning [of the revolt] by their detaining Silius and Velanius; for they thought that they should recover by their means the hostages which they had given to Crassus. The neighbouring people, led on by their influence (as the measures of the Gauls are sudden and hasty), detain Trebius and Terrasidius for the same motive; and quickly sending ambassadors, by means of their leading men, they enter into a mutual compact to do nothing except by general consent, and abide the same issue of fortune; and they solicit the other states to choose rather to continue in that liberty which they had received from their ancestors, than endure slavery under the

Romans. All the sea coast being quickly brought over to their sentiments, they send a common embassy to P. Crassus [to say], "If he wished to receive back his officers, let him send back to them their hostages."

IX.— Caesar, being informed of these things by Crassus, since he was so far distant himself, orders ships of war to be built in the meantime on the river Loire, which flows into the ocean; rowers to be raised from the province; sailors and pilots to be provided. These matters being quickly executed, he himself, as soon as the season of the year permits, hastens to the army. The Veneti, and the other states also, being informed of Caesar's arrival, when they reflected how great a crime they had committed, in that the ambassadors (a character which had amongst all nations ever been sacred and inviolable) had by them been detained and thrown into prison, resolve to prepare for a war in proportion to the greatness of their danger, and especially to provide those things which appertain to the service of a navy; with the greater confidence, inasmuch as they greatly relied on the nature of their situation. They knew that the passes by land were cut off by estuaries, that the approach by sea was most difficult, by reason of our ignorance of the localities, [and] the small number of the harbours, and they trusted that our army would not be able to stay very long among them, on account of the insufficiency of corn; and again, even if all these things should turn out contrary to their expectation, yet they were very powerful in their navy. They, well understood that the Romans neither had any number of ships, nor were acquainted with the shallows, the harbours, or the islands of those parts where they would have to carry on the war; and that navigation was very different in a narrow sea from what it was in the vast and open ocean. Having come to this resolution, they fortify their towns, convey corn into them from the country parts, bring together as many ships as possible to Venetia, where it appeared Caesar would at first carry on the war. They unite to themselves as allies for that war, the Osismii, the Lexovii, the Nannetes, the Ambiliati, the Morini, the Diablintes, and the Menapii; and send for auxiliaries from Britain, which is situated over against those regions.

X.—There were these difficulties which we have mentioned above, in carrying on the war, but many things, nevertheless, urged Caesar to that war; the open insult offered to the state in the detention of the Roman knights, the rebellion raised after surrendering, the revolt after hostages were given, the confederacy of so many states, but principally, lest if [the conduct of] this part was overlooked, the other nations should think that the same thing was permitted them. Wherefore, since he reflected that almost all the Gauls were fond of revolution, and easily and quickly excited to war; that all men likewise, by nature, love liberty and hate the condition of slavery, he thought he ought to divide and more widely distribute his army, before more states should join the confederation.

XI.—He therefore sends T. Labienus, his lieutenant, with the cavalry to the Treviri, who are nearest to the river Rhine. He charges him to visit the Remi and the other Belgians, and to keep them in their allegiance and repel the Germans (who were said to have been summoned by the Belgae to their aid) if they attempted to cross the river by force in their ships. He orders P. Crassus to proceed into Aquitania with twelve legionary cohorts and a great number of the cavalry, lest auxiliaries should be sent into Gaul by these states, and such great nations be united. He sends Q. Titurius Sabinus, his lieutenant, with three legions, among the Unelli, the Curiosolitae, and the Lexovii, to take care that their forces should be kept separate from the rest. He appoints D. Brutus, a young man, over the fleet and those Gallic vessels which he had ordered to be furnished by the Pictones and the Santoni, and the other provinces which remained at peace; and commands him to proceed towards the Veneti, as soon as he could. He himself hastens thither with the land forces.

XII.—The sites of their towns were generally such that, being placed on extreme points [of land] and on promontories, they neither had an approach by land when the tide had rushed in from the main ocean, which always happens twice in the space of twelve hours; nor by ships, because, upon the tide ebbing again, the ships were likely to be dashed upon the shoals. Thus, by either circumstance, was the storming of their towns rendered difficult; and if at any time perchance the Veneti, overpowered by the greatness of our works (the sea having been excluded by a mound and large dams, and the latter being made almost equal in height to the walls of the town), had begun to despair of their fortunes, bringing up a large number of ships, of which they had a very great quantity, they carried off all their property and betook themselves to the nearest towns; there they again defended themselves by the same advantages of situation. They did this the more easily during a great part of the summer, because our ships were kept back by storms, and the difficulty of sailing was very great in that vast and open sea, with its strong tides and its harbours far apart and exceedingly few in number.

XIII.—For their ships were built and equipped after this manner. The keels were somewhat flatter than those of our ships, whereby they could more easily encounter the shallows and the ebbing of the tide: the prows were raised very high, and in like manner the sterns were adapted to the force of the waves and storms [which they were formed to sustain]. The ships were built wholly of oak, and designed to endure any force and violence whatever; the benches, which were made of planks a foot in breadth, were fastened by iron spikes of the thickness of a man's thumb; the anchors were secured fast by iron chains instead of cables, and for sails they used skins and thin dressed leather. These [were used] either through their want of canvas and their ignorance of its application, or for this reason, which is more probable, that they thought that such storms of the ocean, and such violent gales of wind could not be resisted by sails, nor ships of such great burden be conveniently enough managed by them. The encounter of our fleet with these ships was of such a nature that our fleet excelled in speed alone, and the plying of the oars; other things, considering the nature of the place [and] the violence of the storms, were more suitable and better adapted on their side; for neither could our ships injure theirs with their beaks (so great was their strength), nor on account of their height was a weapon easily cast up to them; and for the same reason they were less readily locked in by rocks. To this was added, that whenever a storm began to rage and they ran before the wind, they both could weather the storm more easily and heave to securely in the shallows, and when left by the tide feared nothing from rocks and shelves: the risk of all which things was much to be dreaded by our ships.

XIV. — Caesar, after taking many of their towns, perceiving that so much labour was spent in vain and that the flight of the enemy could not be prevented on the capture of their towns, and that injury could not be done them, he determined to wait for his fleet. As soon as it came up and was first seen by the enemy, about 220 of their ships, fully equipped and appointed with every kind of [naval] implement, sailed forth from the harbour, and drew up opposite to ours; nor did it appear clear to Brutus, who commanded the fleet, or to the tribunes of the soldiers and the centurions, to whom the several ships were assigned, what to do, or what system of tactics to adopt; for they knew that damage could not be done by their beaks; and that, although turrets were built [on their decks], yet the height of the stems of the barbarian ships exceeded these; so that weapons could not be cast up from [our] lower position with sufficient effect, and those cast by the Gauls fell the more forcibly upon us. One thing provided by our men was of great service, [viz.] sharp hooks inserted into and fastened upon poles, of a form not unlike the hooks used in attacking town walls. When the ropes which fastened the sail-yards to the masts were caught by them and pulled, and our vessel vigorously impelled with the oars, they [the ropes] were severed; and when they were cut away, the yards necessarily fell down; so that as all the hope of the Gallic vessels depended on their sails and rigging, upon these being cut away, the entire management of the ships was taken from them at the same time. The rest of the contest depended on courage; in which our men decidedly had the advantage; and the more so because the whole action was carried on in the sight of Caesar and the entire army; so that no act, a little more valiant than ordinary, could pass unobserved, for all the hills and higher grounds, from which there was a near prospect of the sea, were occupied by our army.

XV. — The sail-yards [of the enemy], as we have said, being brought down, although two and [in some cases] three ships [of theirs] surrounded each one [of ours], the soldiers strove with the greatest energy to board the ships of the enemy: and, after the barbarians observed this taking place, as a great many of their ships were beaten, and as no relief for that evil could be discovered, they hastened to seek safety in flight. And, having now turned their vessels to that quarter in which the wind blew, so great a calm and lull suddenly arose, that they could not move out of their place, which circumstance, truly, was exceedingly opportune for finishing the business; for our men gave chase and took them one by one, so that very few out of all the number, [and those] by the intervention of night, arrived at the land, after the battle had lasted almost from the fourth hour till sunset.

XVI. — By this battle the war with the Veneti and the whole of the sea coast was finished; for both all the youth, and all, too, of more advanced age, in whom there was any discretion or rank, had assembled in that battle; and they had collected in that one place whatever naval forces they had anywhere; and when these were lost, the survivors had no place to retreat to, nor means of defending their towns. They accordingly surrendered themselves and all their possessions to Caesar, on whom Caesar thought that punishment should be inflicted the more severely, in order that for the future the rights of ambassadors might be more carefully respected, by barbarians: having, therefore, put to death all their senate, he sold the rest for slaves.

XVII. — While these things are going on amongst the Veneti, Q. Titurius Sabinus with those troops which he had received from Caesar, arrives in the territories of the Unelli. Over these people Viridovix ruled, and held the chief command of all those states which had revolted: from which he had collected a large and powerful army. And in those few days, the Aulerci and the Sexovii, having slain their senate because they would not consent to be promoters of the war, shut their gates [against us] and united themselves to Viridovix; a great multitude besides of desperate men and robbers assembled out of Gaul from all quarters, whom the hope of plundering and the love of fighting had called away from husbandry and their daily labour. Sabinus kept himself within his camp, which was in a position convenient for everything; while Viridovix encamped over against him at a distance of two miles, and daily bringing out his forces, gave him an opportunity of fighting; so that Sabinus had now not only come into contempt with the enemy, but also was somewhat taunted by the speeches of our soldiers; and furnished so great a suspicion of his cowardice that the enemy presumed to approach even to the very rampart of our camp. He adopted this conduct for the following reason: because he did not think that a lieutenant ought to engage in battle with so great a force, especially while he who held the chief command was absent, except on advantageous ground or some favourable circumstance presented itself.

XVIII. — After having established this suspicion of his cowardice, he selected a certain suitable and crafty Gaul, who was one of those whom he had with him as auxiliaries. He induces him by great gifts and promises to go over to the enemy; and informs [him] of what he wished to be done. Who, when he arrives amongst them as a deserter, lays before them the fears of the Romans; and informs them by what difficulties Caesar himself was harassed, and that the matter was not far removed from this — that Sabinus would the next night privately draw off his army out of the camp and set forth to Caesar, for the purpose of carrying [him] assistance, which, when they heard, they all cry out together that an opportunity of successfully conducting their enterprise ought not to be thrown away; that they ought to go to the [Roman] camp. Many things persuaded the Gauls to this measure; the delay of Sabinus during the previous days; the positive assertion of the [pretended] deserter; want of provisions, for a supply of which they had not taken the requisite precautions; the hope springing from the Venetic war; and [also] because in most cases men willingly believe what they wish. Influenced by these things, they do not discharge Viridovix and the other leaders from the council, before they gained permission from them to take up arms and hasten to [our] camp; which being granted, rejoicing as if victory were fully certain, they collected faggots and brushwood, with which to fill up the Roman trenches, and hasten to the camp.

XIX. — The situation of the camp was a rising ground, gently sloping from the bottom for about a mile. Thither they proceeded with great speed (in order that as little time as possible might be given to the Romans to collect and arm themselves), and arrived quite out of breath. Sabinus having encouraged his men, gives them the signal, which they earnestly desired. While the enemy were encumbered by reason of the burdens which they were carrying, he orders a sally to be suddenly made from two gates [of the camp]. It happened, by the advantage of situation, by the unskilfulness and the fatigue of the enemy, by the valour of our soldiers, and their experience in former battles, that they could not stand one attack of our men, and immediately turned their backs: and our men with full vigour followed them while disordered, and slew a great number of them; the horse pursuing the rest, left but few, who escaped by flight. Thus at the same time, Sabinus was informed of the naval battle and Caesar of victory gained by Sabinus; and all the states immediately surrendered themselves to Titurius: for as the temper of the Gauls is impetuous and ready to undertake wars, so their mind is weak, and by no means resolute in enduring calamities.

XX. — About the same time, P. Crassus, when he had arrived in Aquitania (which, as has been before said, both from its extent of territory and the great number of its people, is to be reckoned a third part of Gaul), understanding that he was to wage war in these parts, where a few years before L. Valerius Praeconinus, the lieutenant, had been killed, and his army routed, and from which L. Manilius, the proconsul, had fled with the loss of his baggage, he perceived that no ordinary care must be used by him. Wherefore, having provided corn, procured auxiliaries and cavalry, [and] having summoned by name many valiant men from Tolosa, Carcaso, and Narbo, which are the states of the province of Gaul, that border on these regions [Aquitania], he led his army into the territories of the Sotiates. On his arrival being known, the Sotiates having brought together great forces and [much] cavalry, in which their strength principally lay, and assailing our army on the march, engaged first in a cavalry action, then when their cavalry was routed, and our men pursuing, they suddenly display their infantry forces, which they had placed in ambuscade in a valley. These attacked our men [while] disordered, and renewed the fight.

XXI. — The battle was long and vigorously contested, since the Sotiates, relying on their former victories, imagined that the safety of the whole of Aquitania rested on their valour; [and] our men, on the other hand, desired it might be seen what they could accomplish without their general and without the other legions, under a very young commander; at length the enemy, worn out with wounds, began to turn their backs, and a great number of them being slain, Crassus began to besiege the [principal] town of the Sotiates on his march. Upon their valiantly resisting, he raised vineae and turrets. They at one time attempting a sally, at another forming mines to our rampart and vineae (at which the Aquitani are eminently skilled, because in many places amongst them there are copper mines); when they perceived that nothing could be gained by these operations through the perseverance of our men, they send ambassadors to Crassus, and entreat him to admit them to a surrender. Having obtained it, they, being ordered to deliver up their arms, comply.

XXII. — And while the attention of our men is engaged in that matter, in another part Adcantuannus, who held the chief command, with 600 devoted followers, whom they call soldurii (the conditions of whose association are these, — that they enjoy all the conveniences of life with those to whose friendship they have devoted themselves: if anything calamitous happen to them, either they endure the same destiny together with them, or commit suicide: nor hitherto, in the memory of men, has there been found any one who, upon his being slain to whose friendship he had devoted himself, refused to die); Adcantuannus, [I say] endeavouring to make a sally with these, when our soldiers had rushed together to arms, upon a shout being raised at that part of the fortification, and a fierce battle had been fought there, was driven back into the town, yet he obtained from Crassus [the indulgence] that he should enjoy the same terms of surrender [as the other inhabitants].

XXIII. — Crassus, having received their arms and hostages, marched into the territories of the Vocates and the Tarusates. But then, the barbarians being alarmed, because they had heard that a town fortified by the nature of the place and by art had been taken by us in a few days after our arrival there, began to send ambassadors into all quarters, to combine, to give hostages one to another, to raise troops. Ambassadors also are sent to those states of Hither Spain which are nearest to Aquitania, and auxiliaries and leaders are summoned from them; on whose arrival they proceed to carry on the war with great confidence, and with a great host of men. They who had been with Q. Sertorius the whole period [of his war in Spain] and were supposed to have very great skill in military matters, are chosen leaders. These, adopting the practice of the Roman people, begin to select [advantageous] places, to fortify their camp, to cut off our men from provisions, which, when Crassus observes, [and likewise] that his forces, on account of their small number, could not safely be separated; that the enemy both made excursions and beset the passes, and [yet] left sufficient guard for their camp; that on that account, corn and provision could not very conveniently be brought up to him, and that the number of the enemy was daily increased, he thought that he ought not to delay in giving battle. This matter being brought to a council, when he discovered that all thought the same thing, he appointed the next day for the fight.

XXIV. — Having drawn out all his forces at the break of day, and marshalled them in a double line, he posted the auxiliaries in the centre, and waited to see what measures the enemy would take. They, although on account of their great number and their ancient renown in war, and the small number of our men, they supposed they might safely fight, nevertheless considered it safer to gain the victory without any wound, by besetting the passes [and] cutting off the provisions: and if the Romans, on account of the want of corn, should begin to retreat, they intended to attack them while encumbered in their march and depressed in spirit [as being assailed while] under baggage. This measure being approved of by the leaders and the forces of the Romans drawn out, the enemy [still] kept themselves in their camp. Crassus having remarked this circumstance, since the enemy, intimidated by their own delay, and by the reputation [i.e. for cowardice arising thence] had rendered our soldiers more eager for fighting, and the remarks of all were heard [declaring] that no longer ought delay to be made in going to the camp, after encouraging his men, he marches to the camp of the enemy, to the great gratification of his own troops.

XXV. — There, while some were filling up the ditch, and others, by throwing a large number of darts, were driving the defenders from the rampart and fortifications, and the auxiliaries, on whom Crassus did not much rely in the battle, by supplying stones and weapons [to the soldiers], and by conveying turf to the mound, presented the appearance and character of men engaged in fighting; while also the enemy were fighting resolutely and boldly, and their weapons, discharged from their higher position, fell with great effect; the horse, having gone round the camp of the enemy, reported to Crassus that the camp was not fortified with equal care on the side of the Decuman gate, and had an easy approach.

XXVI. — Crassus, having exhorted the commanders of the horse to animate their men by great rewards and promises, points out to them what he wished to have done. They, as they had been commanded, having brought out the four cohorts, which, as they had been left as a guard for the camp, were not fatigued by exertion, and having led them round by a somewhat longer way, lest they could be seen from the camp of the enemy, when the eyes and minds of all were intent upon the battle, quickly arrived at those fortifications which we have spoken of, and, having demolished these, stood in the camp of the enemy before they were seen by them, or it was known what was going on. And then, a shout being heard in that quarter, our men, their strength having been recruited (which usually occurs on the hope of victory), began to fight more vigorously. The enemy, surrounded on all sides, [and] all their affairs being despaired of, made great attempts to cast themselves down over the ramparts and to seek safety in flight. These the cavalry pursued over the very open plains, and after leaving scarcely a fourth part out of the number of 50,000, which it was certain had assembled out of Aquitania and from the Cantabri, returned late at night to the camp.

XXVII. — Having heard of this battle, the greatest part of Aquitania surrendered itself to Crassus, and of its own accord sent hostages, in which number were the Tarbelli, the Bigerriones, the Preciani, the Vocasates, the Tarusates, the Elurates, the Garites, the Ausci, the Garumni, the Sibuzates, the Cocosates. A few [and those] most remote nations, relying on the time of the year, because winter was at hand, neglected to do this.

XXVIII. — About the same time Caesar, although the summer was nearly past, yet since, all Gaul being reduced, the Morini and the Menapii alone remained in arms, and had never sent ambassadors to him [to make a treaty] of peace, speedily led his army thither, thinking that that war might soon be terminated. They resolved to conduct the war on a very different method from the rest of the Gauls; for as they perceived that the greatest nations [of Gaul] who had engaged in war, had been routed and overcome, and as they possessed continuous ranges of forests and morasses, they removed themselves and all their property thither. When Caesar had arrived at the opening of these forests, and had begun to fortify his camp, and no enemy was in the meantime seen, while our men were dispersed on their respective duties, they suddenly rushed out from all parts of the forest, and made an attack on our men. The latter quickly took up arms and drove them back again to their forests; and having killed a great many, lost a few of their own men while pursuing them too far through those intricate places.

XXIX. — During the remaining days after this, Caesar began to cut down the forests; and that no attack might be made on the flank of the soldiers, while unarmed and not foreseeing it, he placed together (opposite to the enemy) all that timber which was cut down, and piled it up as a rampart on either flank. When a great space had been, with incredible speed, cleared in a few days, when the cattle [of the enemy] and the rear of their baggage-train were already seized by our men, and they themselves were seeking for the thickest parts of the forests, storms of such a kind came on that the work was necessarily suspended, and, through the continuance of the rains, the soldiers could not any longer remain in their tents. Therefore, having laid waste all their country, [and] having burnt their villages and houses, Caesar led back his army and stationed them in winter-quarters among the Aulerci and Lexovii, and the other states which had made war upon him last.

Book IV

I. — The following winter (this was the year in which Cn. Pompey and M. Crassus were consuls), those Germans [called] the Usipetes, and likewise the Tenchtheri, with a great number of men, crossed the Rhine, not far from the place at which that river discharges itself into the sea. The motive for crossing [that river] was that, having been for several years harassed by the Suevi, they were constantly engaged in war, and hindered from the pursuits of agriculture. The nation of the Suevi is by far the largest and the most warlike nation of all the Germans. They are said to possess a hundred cantons, from each of which they yearly send from their territories for the purpose of war a thousand armed men: the others who remain at home, maintain [both] themselves and those engaged in the expedition. The latter again, in their turn, are in arms the year after: the former remain at home. Thus neither husbandry nor the art and practice of war are neglected. But among them there exists no private and separate land; nor are they permitted to remain more than one year in one place for the purpose of residence. They do not live much on corn, but subsist for the most part on milk and flesh, and are much [engaged] in hunting; which circumstance must, by the nature of their food, and by their daily exercise and the freedom of their life (for having from boyhood been accustomed to no employment, or discipline, they do nothing at all contrary to their inclination), both promote their strength and render them men of vast stature of body. And to such a habit have they brought themselves, that even in the coldest parts they wear no clothing whatever except skins, by reason of the scantiness of which a great portion of their body is bare, and besides they bathe in open rivers.

II. — Merchants have access to them rather that they may have persons to whom they may sell those things which they have taken in war, than because they need any commodity to be imported to them. Moreover, even as to labouring cattle, in which the Gauls take the greatest pleasure, and which they procure at a great price, the Germans do not employ such as are imported, but those poor and ill-shaped animals which belong to their country; these, however, they render capable of the greatest labour by daily exercise. In cavalry actions they frequently leap from their horses and fight on foot; and train their horses to stand still in the very spot on which they leave them, to which they retreat with great activity when there is occasion; nor, according to their practice, is anything regarded as more unseemly, or more unmanly, than to use housings. Accordingly, they have the courage, though they be themselves but few, to advance against any number whatever of horse mounted with housings. They on no account permit wine to be imported to them, because they consider that men degenerate in their powers of enduring fatigue, and are rendered effeminate by that commodity.

III. — They esteem it their greatest praise as a nation that the lands about their territories lie unoccupied to a very great extent, inasmuch as [they think] that by this circumstance is indicated that a great number of nations cannot, withstand their power; and thus on one side of the Suevi the lands are said to lie desolate for about six hundred miles. On the other side they border on the Ubii, whose state was large and flourishing, considering the condition of the Germans, and who are somewhat more refined than those of the same race and the rest [of the Germans], and that because they border on the Rhine, and are much resorted to by merchants, and are accustomed to the manners of the Gauls, by reason of their approximity to them. Though the Suevi, after making the attempt frequently and in several wars, could not expel this nation from their territories, on account of the extent and population of their state, yet they made them tributaries, and rendered them less distinguished and powerful [than they had ever been].

IV. — In the same condition were the Usipetes and the Tenchtheri (whom we have mentioned above), who for many years resisted the power of the Suevi, but being at last driven from their possessions, and having wandered through many parts of Germany, came to the Rhine, to districts which the Menapii inhabited, and where they had lands, houses, and villages on either side of the river. The latter people, alarmed by the arrival of so great a multitude, removed from those houses which they had on the other side of the river, and having placed guards on this side the Rhine, proceeded to hinder the Germans from crossing. They, finding themselves, after they had tried all means, unable either to force a passage on account of their deficiency in shipping, or cross by stealth on account of the guards of the Menapii, pretended to return to their own settlements and districts; and, after having proceeded three days' march, returned; and their cavalry having performed the whole of this journey in one night, cut off the Menapii, who were ignorant of, and did not expect [their approach, and] who, having moreover been informed of the departure of the Germans by their scouts, had without apprehension returned to their villages beyond the Rhine. Having slain these, and seized their ships, they crossed the river before that part of the Menapii, who were at peace in their settlements over the Rhine, were apprised of [their intention]; and seizing all their houses, maintained themselves upon their provisions during the rest of the winter.

V. — Caesar, when informed of these matters, fearing the fickle disposition of the Gauls, who are easily prompted to take up resolutions, and much addicted to change, considered that nothing was to be entrusted to them; for it is the custom of that people to compel travellers to stop, even against their inclination, and inquire what they may have heard, or may know, respecting any matter; and in towns the common people throng around merchants and force them to state from what countries they come, and what affairs they know of there. They often engage in resolutions concerning the most important matters, induced by these reports and stories alone; of which they must necessarily instantly repent, since they yield to mere unauthorised reports; and since most people give to their questions answers framed agreeably to their wishes.

VI. — Caesar, being aware of their custom, in order that he might not encounter a more formidable war, sets forward to the army earlier in the year than he was accustomed to do. When he had arrived there, he discovered that those things, which he had suspected would occur, had taken place; that embassies had been sent to the Germans by some of the states, and that they had been entreated to leave the Rhine, and had been promised that all things which they desired should be provided by the Gauls. Allured by this hope, the Germans were then making excursions to greater distances, and had advanced to the territories of the Eburones and the Condrusi, who are under the protection of the Treviri. After summoning the chiefs of Gaul, Caesar thought proper to pretend ignorance of the things which he had discovered; and having conciliated and confirmed their minds, and ordered some cavalry to be raised, resolved to make war against the Germans.

VII. — Having provided corn and selected his cavalry, he began to direct his march towards those parts in which he heard the Germans were. When he was distant from them only a few days' march, ambassadors come to him from their state; whose speech was as follows: — "That the Germans neither make war upon the Roman people first, nor do they decline, if they are provoked, to engage with them in arms; for that this was the custom of the Germans handed down to them from their forefathers, to resist whatsoever people make war upon them and not to avert it by entreaty; this, however, they confessed, — that they had come hither reluctantly, having been expelled from their country. If the Romans were disposed to accept their friendship, they might be serviceable allies to them; and let them either assign them lands, or permit them to retain those which they had acquired by their arms; that they are inferior to the Suevi alone, to whom not even the immortal gods can show themselves equal; that there was none at all besides on earth whom they could not conquer."

VIII. — To these remarks Caesar replied in such terms as he thought proper; but the conclusion of his speech was, "That he could make no alliance with them, if they continued in Gaul; that it was not probable that they who were not able to defend their own territories, should get possession of those of others, nor were there any lands lying waste in Gaul which could be given away, especially to so great a number of men, without doing wrong [to others]; but they might, if they were desirous, settle in the territories of the Ubii; whose ambassadors were then with him, and were complaining of the aggressions of the Suevi, and requesting assistance from him; and that he would obtain this request from them."

IX. — The ambassadors said that they would report these things to their countrymen; and, after having deliberated on the matter, would return to Caesar after the third day, they begged that he would not in the meantime advance his camp nearer to them. Caesar said that he could not grant them even that; for he had learned that they had sent a great part of their cavalry over the Meuse to the Ambivariti, some days before, for the purpose of plundering and procuring forage. He supposed that they were then waiting for these horse, and that the delay was caused on this account.

X. — The Meuse rises from mount Le Vosge, which is in the territories of the Lingones; and, having received a branch of the Rhine, which is called the Waal, forms the island of the Batavi, and not more than eighty miles from it it falls into the ocean. But the Rhine takes its course among the Lepontii, who inhabit the Alps, and is carried with a rapid current for a long distance through the territories of the Sarunates, Helvetii, Sequani, Mediomatrici, Tribuci, and Treviri, and when it approaches the ocean, divides into several branches; and, having formed many and extensive islands, a great part of which are inhabited by savage and barbarous nations (of whom there are some who are supposed to live on fish and the eggs of sea-fowl), flows into the ocean by several mouths.

XI. — When Caesar was not more than twelve miles distant from the enemy, the ambassadors return to him, as had been arranged; who meeting him on the march, earnestly entreated him not to advance any farther. When they could not obtain this, they begged him to send on a despatch to those who had marched in advance of the main army, and forbid them to engage; and grant them permission to send ambassadors to the Ubii, and if the princes and senate of the latter would give them security by oath, they assured Caesar that they would accept such conditions as might be proposed by him; and requested that he would give them the space of three days for negotiating these affairs. Caesar thought that these things tended to the self-same point [as their other proposal]; [namely] that, in consequence of a delay of three days intervening, their horse which were at a distance might return; however, he said, that he would not that day advance farther than four miles for the purpose of procuring water; he ordered that they should assemble at that place in as large a number as possible the following day, that he might inquire into their demands. In the meantime he sends messengers to the officers who had marched in advance with all the cavalry to order them not to provoke the enemy to an engagement, and if they themselves were assailed, to sustain the attack until he came up with the army.

XII. — But the enemy, as soon as they saw our horse, the number of which was 5,000, whereas they themselves had not more than 800 horse, because those which had gone over the Meuse for the purpose of foraging had not returned, while our men had no apprehensions, because their ambassadors had gone away from Caesar a little before, and that day had been requested by them as a period of truce, made an onset on our men, and soon threw them into disorder. When our men, in their turn, made a stand, they, according to their practice, leaped from their horses to their feet, and stabbing our horses in the belly and overthrowing a great many of our men, put the rest to flight, and drove them forward so much alarmed that they did not desist from their retreat till they had come in sight of our army. In that encounter seventy-four of our horse were slain; among them, Piso, an Aquitanian, a most valiant man, and descended from a very illustrious family; whose grandfather had held the sovereignty of his state, and had been styled friend by our senate. He, while he was endeavouring to render assistance to his brother who was surrounded by the enemy, and whom he rescued from danger, was himself thrown from his horse, which was wounded under him, but still opposed [his antagonists] with the greatest intrepidity, as long as he was able to maintain the conflict. When at length he fell, surrounded on all sides and after receiving many wounds, and his brother, who had then retired from the fight, observed it from a distance, he spurred on his horse, threw himself upon the enemy, and was killed.

XIII. — After this engagement, Caesar considered that neither ought ambassadors to be received to audience, nor conditions be accepted by him from those who, after having sued for peace by way of stratagem and treachery, had made war without provocation. And to wait till the enemy's forces were augmented and their cavalry had returned, he concluded, would be the greatest madness; and knowing the fickleness of the Gauls, he felt how much influence the enemy had already acquired among them by this one skirmish. He [therefore] deemed that no time for converting measures ought to be afforded them. After having resolved on these things and communicated his plans to his lieutenants and quaestor in order that he might not suffer any opportunity for engaging to escape him, a very seasonable event occurred, namely, that on the morning of the next day, a large body of Germans, consisting of their princes and old men, came to the camp to him to practise the same treachery and dissimulation; but, as they asserted, for the purpose of acquitting themselves for having engaged in a skirmish the day before, contrary to what had been agreed and to what, indeed, they themselves had requested; and also if they could by any means obtain a truce by deceiving him. Caesar, rejoicing that they had fallen into his power, ordered them to be detained. He then drew all his forces out of the camp, and commanded the cavalry, because he thought they were intimidated by the late skirmish, to follow in the rear.

XIV. — Having marshalled his army in three lines, and in a short time performed a march of eight miles, he arrived at the camp of the enemy before the Germans could perceive what was going on; who being suddenly alarmed by all the circumstances, both by the speediness of our arrival and the absence of their own officers, as time was afforded neither for concerting measures nor for seizing their arms, are perplexed as to whether it would be better to lead out their forces against the enemy, or to defend their camp, or seek their safety by flight. Their consternation being made apparent by their noise and tumult, our soldiers, excited by the treachery of the preceding day, rushed into the camp: such of them as could readily get their arms for a short time withstood our men, and gave battle among their carts and baggage-waggons; but the rest of the people, [consisting] of boys and women (for they had left their country and crossed the Rhine with all their families), began to fly in all directions; in pursuit of whom Caesar sent the cavalry.

XV. — The Germans when, upon hearing a noise behind them, [they looked and] saw that their families were being slain, throwing away their arms and abandoning their standards, fled out of the camp, and when they had arrived at the confluence of the Meuse and the Rhine, the survivors despairing of farther escape, as a great number of their countrymen had been killed, threw themselves into the river and there perished, overcome by fear, fatigue, and the violence of the stream. Our soldiers, after the alarm of so great a war, for the number of the enemy amounted to 430,000, returned to their camp, all safe to a man, very few being even wounded. Caesar granted those whom he had detained in the camp liberty of departing. They however, dreading revenge and torture from the Gauls, whose lands they had harassed, said that they desired to remain with him. Caesar granted them permission.

XVI. — The German war being finished, Caesar thought it expedient for him to cross the Rhine, for many reasons; of which this was the most weighty, that, since he saw the Germans were so easily urged to go into Gaul, he desired they should have their fears for their own territories when they discovered that the army of the Roman people both could and dared pass the Rhine. There was added also, that that portion of the cavalry of the Usipetes and the Tenchtheri, which I have above related to have crossed the Meuse for the purpose of plundering and procuring forage, and was not present at the engagement, had betaken themselves, after the retreat of their countrymen, across the Rhine into the territories of the Sigambri, and united themselves to them. When Caesar sent ambassadors to them, to demand that they should give up to him those who had made war against him and against Gaul, they replied, "That the Rhine bounded the empire of the Roman people; if he did not think it just for the Germans to pass over into Gaul against his consent, why did he claim that anything beyond the Rhine should be subject to his dominion or power?" The Ubii also, who alone, out of all the nations lying beyond the Rhine, had sent ambassadors to Caesar, and formed an alliance and given hostages, earnestly entreated "that he would bring them assistance, because they were grievously oppressed by the Suevi; or, if he was prevented from doing so by the business of the commonwealth, he would at least transport his army over the Rhine; that that would be sufficient for their present assistance and their hope for the future; that so great was the name and the reputation of his army, even among the most remote nations of the Germans, arising from the defeat of Ariovistus and this last battle which was fought, that they might be safe under the fame and friendship of the Roman people." They promised a large number of ships for transporting the army.

XVII. — Caesar, for those reasons which I have mentioned, had resolved to cross the Rhine; but to cross by ships he neither deemed to be sufficiently safe, nor considered consistent with his own dignity or that of the Roman people. Therefore, although the greatest difficulty in forming a bridge was presented to him, on account of the breadth, rapidity, and depth of the river, he nevertheless considered that it ought to be attempted by him, or that his army ought not otherwise to be led over. He devised this plan of a bridge. He joined together at the distance of two feet, two piles, each a foot and a half thick, sharpened a little at the lower end, and proportioned in length to the depth of the river. After he had, by means of engines, sunk these into the river, and fixed them at the bottom, and then driven them in with rammers, not quite perpendicularly, like a stake, but bending forward and sloping, so as to incline in the direction of the current of the river; he also placed two [other piles] opposite to these, at the distance of forty feet lower down, fastened together in the same manner, but directed against the force and current of the river. Both these, moreover, were kept firmly apart by beams two feet thick (the space which the binding of the piles occupied), laid in at their extremities between two braces on each side; and in consequence of these being in different directions and fastened on sides the one opposite to the other, so great was the strength of the work, and such the arrangement of the materials, that in proportion as the greater body of water dashed against the bridge, so much the closer were its parts held fastened together. These beams were bound together by timber laid over them in the direction of the length of the bridge, and were [then] covered over with laths and hurdles; and in addition to this, piles were driven into the water obliquely, at the lower side of the bridge, and these serving as buttresses, and being connected with every portion of the work, sustained the force of the stream: and there were others also above the bridge, at a moderate distance; that if trunks of trees or vessels were floated down the river by the barbarians for the purpose of destroying the work, the violence of such things might be diminished by these defences, and might not injure the bridge.

XVIII.— Within ten days after the timber began to be collected, the whole work was completed, and the whole army led over. Caesar, leaving a strong guard at each end of the bridge, hastens into the territories of the Sigambri. In the meantime ambassadors from several nations come to him, whom, on their suing for peace and alliance, he answers in a courteous manner, and orders hostages to be brought to him. But the Sigambri, at the very time the bridge was begun to be built, made preparations for a flight (by the advice of such of the Tenchtheri and Usipetes as they had amongst them), and quitted their territories and conveyed away all their possessions, and concealed themselves in deserts and woods.

XIX. — Caesar, having remained in their territories a few days, and burnt all their villages and houses, and cut down their corn, proceeded into the territories of the Ubii; and having promised them his assistance, if they were ever harassed by the Suevi, he learned from them these particulars: that the Suevi, after they had by means of their scouts found that the bridge was being built, had called a council, according to their custom, and sent orders to all parts of their state to remove from the towns and convey their children, wives, and all their possessions into the woods, and that all who could bear arms should assemble in one place; that the place thus chosen was nearly the centre of those regions which the Suevi possessed; that in this spot they had resolved to await the arrival of the Romans, and give them battle there. When Caesar discovered this, having already accomplished all those things on account of which he had resolved to lead his army over, namely, to strike fear into the Germans, take vengeance on the Sigambri, and free the Ubii from the invasion of the Suevi, having spent altogether eighteen days beyond the Rhine, and thinking he had advanced far enough to serve both honour and interest, he returned into Gaul, and cut down the bridge.

XX. — During the short part of summer which remained, Caesar, although in these countries, as all Gaul lies towards the north, the winters are early, nevertheless resolved to proceed into Britain, because he discovered that in almost all the wars with the Gauls succours had been furnished to our enemy from that country; and even if the time of year should be insufficient for carrying on the war, yet he thought it would be of great service to him if he only entered the island, and saw into the character of the people, and got knowledge of their localities, harbours, and landing-places, all which were for the most part unknown to the Gauls. For neither does any one except merchants generally go thither, nor even to them was any portion of it known, except the sea-coast and those parts which are opposite to Gaul. Therefore, after having called up to him the merchants from all parts, he could learn neither what was the size of the island, nor what or how numerous were the nations which inhabited it, nor what system of war they followed, nor what customs they used, nor what harbours were convenient for a great number of large ships.

XXI. — He sends before him Caius Volusenus with a ship of war, to acquire a knowledge of these particulars before he in person should make a descent into the island, as he was convinced that this was a judicious measure. He commissioned him to thoroughly examine into all matters, and then return to him as soon as possible. He himself proceeds to the Morini with all his forces. He orders ships from all parts of the neighbouring countries, and the fleet which the preceding summer he had built for the war with the Veneti, to assemble in this place. In the meantime, his purpose having been discovered, and reported to the Britons by merchants, ambassadors come to him from several states of the island, to promise that they will give hostages, and submit to the government of the Roman people. Having given them an audience, he after promising liberally, and exhorting them to continue in that purpose, sends them back to their own country, and [despatches] with them Commius, whom, upon subduing the Atrebates, he had created king there, a man whose courage and conduct he esteemed, and who he thought would be faithful to him, and whose influence ranked highly in those countries. He orders him to visit as many states as he could, and persuade them to embrace the protection of the Roman people, and apprise them that he would shortly come thither. Volusenus, having viewed the localities as far as means could be afforded one who dared not leave his ship and trust himself to barbarians, returns to Caesar on the fifth day, and reports what he had there observed.

XXII. — While Caesar remains in these parts for the purpose of procuring ships, ambassadors come to him from a great portion of the Morini, to plead their excuse respecting their conduct on the late occasion; alleging that it was as men uncivilised, and as those who were unacquainted with our custom, that they had made war upon the Roman people, and promising to perform what he should command. Caesar, thinking that this had happened fortunately enough for him, because he neither wished to leave an enemy behind him, nor had an opportunity for carrying on a war, by reason of the time of year, nor considered that employment in such trifling matters was to be preferred to his enterprise on Britain, imposes a large number of hostages; and when these were brought, he received them to his protection. Having collected together and provided about eighty transport ships, as many as he thought necessary for conveying over two legions, he assigned such [ships] of war as he had besides to the quaestor, his lieutenants, and officers of cavalry. There were in addition to these eighteen ships of burden which were prevented, eight miles from that place, by winds, from being able to reach the same port. These he distributed amongst the horse; the rest of the army he delivered to Q. Titurius Sabinus and L. Aurunculeius Cotta, his lieutenants, to lead into the territories of the Menapii and those cantons of the Morini from which ambassadors had not come to him. He ordered P. Sulpicius Rufus, his lieutenant, to hold possession of the harbour, with such a garrison as he thought sufficient.

XXIII. — These matters being arranged, finding the weather favourable for his voyage, he set sail about the third watch, and ordered the horse to march forward to the farther port, and there embark and follow him. As this was performed rather tardily by them, he himself reached Britain with the first squadron of ships, about the fourth hour of the day, and there saw the forces of the enemy drawn up in arms on all the hills. The nature of the place was this: the sea was confined by mountains so close to it that a dart could be thrown from their summit upon the shore. Considering this by no means a fit place for disembarking, he remained at anchor till the ninth hour, for the other ships to arrive there. Having in the meantime assembled the lieutenants and military tribunes, he told them both what he had learnt from Volusenus, and what he wished to be done; and enjoined them (as the principle of military matters, and especially as maritime affairs, which have a precipitate and uncertain action, required) that all things should be performed by them at a nod and at the instant. Having dismissed them, meeting both with wind and tide favourable at the same time, the signal being given and the anchor weighed, he advanced about seven miles from that place, and stationed his fleet over against an open and level shore.

XXIV. — But the barbarians, upon perceiving the design of the Romans, sent forward their cavalry and charioteers, a class of warriors of whom it is their practice to make great use in their battles, and following with the rest of their forces, endeavoured to prevent our men landing. In this was the greatest difficulty, for the following reasons, namely, because our ships, on account of their great size, could be stationed only in deep water; and our soldiers, in places unknown to them, with their hands embarrassed, oppressed with a large and heavy weight of armour, had at the same time to leap from the ships, stand amidst the waves, and encounter the enemy; whereas they, either on dry ground, or advancing a little way into the water, free in all their limbs, in places thoroughly known to them, could confidently throw their weapons and spur on their horses, which were accustomed to this kind of service. Dismayed by these circumstances and altogether untrained in this mode of battle, our men did not all exert the same vigour and eagerness which they had been wont to exert in engagements on dry ground.

XXV. — When Caesar observed this, he ordered the ships of war, the appearance of which was somewhat strange to the barbarians and the motion more ready for service, to be withdrawn a little from the transport vessels, and to be propelled by their oars, and be stationed towards the open flank of the enemy, and the enemy to be beaten off and driven away with slings, arrows, and engines: which plan was of great service to our men; for the barbarians being startled by the form of our ships and the motions of our oars and the nature of our engines, which was strange to them, stopped, and shortly after retreated a little. And while our men were hesitating [whether they should advance to the shore], chiefly on account of the depth of the sea, he who carried the eagle of the tenth legion, after supplicating the gods that the matter might turn out favourably to the legion, exclaimed, "Leap, fellow soldiers, unless you wish to betray your eagle to the enemy. I, for my part, will perform my duty to the commonwealth and my general." When he had said this with a loud voice, he leaped from the ship and proceeded to bear the eagle toward the enemy. Then our men, exhorting one another that so great a disgrace should not be incurred, all leaped from the ship. When those in the nearest vessels saw them, they speedily followed and approached the enemy.

XXVI. — The battle was maintained vigorously on both sides. Our men, however, as they could neither keep their ranks, nor get firm footing, nor follow their standards, and as one from one ship and another from another assembled around whatever standards they met, were thrown into great confusion. But the enemy, who were acquainted with all the shallows, when from the shore they saw any coming from a ship one by one, spurred on their horses, and attacked them while embarrassed; many surrounded a few, others threw their weapons upon our collected forces on their exposed flank. When Caesar observed this, he ordered the boats of the ships of war and the spy sloops to be filled with soldiers, and sent them up to the succour of those whom he had observed in distress. Our men, as soon as they made good their footing on dry ground, and all their comrades had joined them, made an attack upon the enemy, and put them to flight, but could not pursue them very far, because the horse had not been able to maintain their course at sea and reach the island. This alone was wanting to Caesar's accustomed success.

XXVII. — The enemy being thus vanquished in battle, as soon as they recovered after their flight, instantly sent ambassadors to Caesar to negotiate about peace. They promised to give hostages and perform what he should command. Together with these ambassadors came Commius the Atrebatian, who, as I have above said, had been sent by Caesar into Britain. Him they had seized upon when leaving his ship, although in the character of ambassador he bore the general's commission to them, and thrown into chains: then after the battle was fought, they sent him back, and in suing for peace cast the blame of that act upon the common people, and entreated that it might be pardoned on account of their indiscretion. Caesar, complaining that after they had sued for peace, and had voluntarily sent ambassadors into the continent for that purpose, they had made war without a reason, said that he would pardon their indiscretion, and imposed hostages, a part of whom they gave immediately; the rest they said they would give in a few days, since they were sent for from remote places. In the meantime they ordered their people to return to the country parts, and the chiefs assembled from all quarters, and proceeded to surrender themselves and their states to Caesar.

XXVIII. — A peace being established by these proceedings four days after we had come into Britain, the eighteen ships, to which reference has been made above, and which conveyed the cavalry, set sail from the upper port with a gentle gale; when, however, they were approaching Britain and were seen from the camp, so great a storm suddenly arose that none of them could maintain their course at sea; and some were taken back to the same port from which they had started; — others, to their great danger, were driven to the lower part of the island, nearer to the west; which, however, after having cast anchor, as they were getting filled with water, put out to sea through necessity in a stormy night, and made for the continent.

XXIX. — It happened that night to be full moon, which usually occasions very high tides in that ocean; and that circumstance was unknown to our men. Thus, at the same time, the tide began to fill the ships of war which Caesar had provided to convey over his army, and which he had drawn up on the strand; and the storm began to dash the ships of burden which were riding at anchor against each other; nor was any means afforded our men of either managing them or of rendering any service. A great many ships having been wrecked, inasmuch as the rest, having lost their cables, anchors, and other tackling, were unfit for sailing, a great confusion, as would necessarily happen, arose throughout the army; for there were no other ships in which they could be conveyed back, and all things which are of service in repairing vessels were wanting, and corn for the winter had not been provided in those places, because it was understood by all that they would certainly winter in Gaul.

XXX. — On discovering these things the chiefs of Britain, who had come up after the battle was fought to perform those conditions which Caesar had imposed, held a conference, when they perceived that cavalry, and ships, and corn were wanting to the Romans, and discovered the small number of our soldiers from the small extent of the camp (which, too, was on this account more limited than ordinary because Caesar had conveyed over his legions without baggage), and thought that the best plan was to renew the war, and cut off our men from corn and provisions and protract the affair till winter; because they felt confident that, if they were vanquished or cut off from a return, no one would afterwards pass over into Britain for the purpose of making war. Therefore, again entering into a conspiracy, they began to depart from the camp by degrees and secretly bring up their people from the country parts.

XXXI. — But Caesar, although he had not as yet discovered their measures, yet, both from what had occurred to his ships, and from the circumstance that they had neglected to give the promised hostages, suspected that the thing would come to pass which really did happen. He therefore provided remedies against all contingencies; for he daily conveyed corn from the country parts into the camp, used the timber and brass of such ships as were most seriously damaged for repairing the rest, and ordered whatever things besides were necessary for this object to be brought to him from the continent. And thus, since that business was executed by the soldiers with the greatest energy, he effected that, after the loss of twelve ships, a voyage could be made well enough in the rest.

XXXII. — While these things are being transacted, one legion had been sent to forage, according to custom, and no suspicion of war had arisen as yet, and some of the people remained in the country parts, others went backwards and forwards to the camp, they who were on duty at the gates of the camp reported to Caesar that a greater dust than was usual was seen in that direction in which the legion had marched. Caesar, suspecting that which was [really the case], — that some new enterprise was undertaken by the barbarians, ordered the two cohorts which were on duty to march into that quarter with him, and two other cohorts to relieve them on duty; the rest to be armed and follow him immediately. When he had advanced some little way from the camp, he saw that his men were overpowered by the enemy and scarcely able to stand their ground, and that, the legion being crowded together, weapons were being cast on them from all sides. For as all the corn was reaped in every part with the exception of one, the enemy, suspecting that our men would repair to that, had concealed themselves in the woods during the night. Then attacking them suddenly, scattered as they were, and when they had laid aside their arms, and were engaged in reaping, they killed a small number, threw the rest into confusion, and surrounded them with their cavalry and chariots.

XXXIII. — Their mode of fighting with their chariots is this: firstly, they drive about in all directions and throw their weapons and generally break the ranks of the enemy with the very dread of their horses and the noise of their wheels; and when they have worked themselves in between the troops of horse, leap from their chariots and engage on foot. The charioteers in the meantime withdraw some little distance from the battle, and so place themselves with the chariots that, if their masters are overpowered by the number of the enemy, they may have a ready retreat to their own troops. Thus they display in battle the speed of horse, [together with] the firmness of infantry; and by daily practice and exercise attain to such expertness that they are accustomed, even on a declining and steep place, to check their horses at full speed, and manage and turn them in an instant and run along the pole, and stand on the yoke, and thence betake themselves with the greatest celerity to their chariots again.

XXXIV. —Under these circumstances, our men being dismayed by the novelty of this mode of battle, Caesar most seasonably brought assistance; for upon his arrival the enemy paused, and our men recovered from their fear; upon which, thinking the time unfavourable for provoking the enemy and coming to an action, he kept himself in his own quarter, and, a short time having intervened, drew back the legions into the camp. While these things were going on, and all our men engaged, the rest of the Britons, who were in the fields, departed. Storms then set in for several successive days, which both confined our men to camp and hindered the enemy from attacking us. In the meantime the barbarians despatched messengers to all parts and reported to their people the small number of our soldiers, and how good an opportunity was given for obtaining spoil and for liberating themselves for ever, if they should only drive the Romans from their camp. Having by these means speedily got together a large force of infantry and of cavalry, they came up to the camp.

XXXV. — Although Caesar anticipated that the same thing which had happened on former occasions would then occur — that, if the enemy were routed, they would escape from danger by their speed; still, having got about thirty horse, which Commius the Atrebatian, of whom mention has been made, had brought over with him [from Gaul], he drew up the legions in order of battle before the camp. When the action commenced, the enemy were unable to sustain the attack of our men long, and turned their backs; our men pursued them as far as their speed and strength permitted, and slew a great number of them; then, having destroyed and burnt everything far and wide, they retreated to their camp.

XXXVI. — The same day, ambassadors sent by the enemy came to Caesar to negotiate a peace. Caesar doubled the number of hostages which he had before demanded; and ordered that they should be brought over to the continent, because, since the time of the equinox was near, he did not consider that, with his ships out of repair, the voyage ought to be deferred till winter. Having met with favourable weather he set sail a little after midnight, and all his fleet arrived safe at the continent, except two of the ships of burden which could not make the same port which the other ships did, and were carried a little lower down.

XXXVII. — When our soldiers, about 300 in number, had been drawn out of these two ships, and were marching to the camp, the Morini, whom Caesar, when setting forth for Britain, had left in a state of peace, excited by the hope of spoil, at first surrounded them with a small number of men, and ordered them to lay down their arms, if they did not wish to be slain; afterwards however, when they, forming a circle, stood on their defence, a shout was raised and about 6,000 of the enemy soon assembled; which being reported, Caesar sent all the cavalry in the camp as a relief to his men. In the meantime our soldiers sustained the attack of the enemy, and fought most valiantly for more than four hours, and, receiving but few wounds themselves, slew several of them. But after our cavalry came in sight, the enemy, throwing away their arms, turned their backs, and a great number of them were killed.

XXXVIII. — The day following Caesar sent Labienus, his lieutenant, with those legions which he had brought back from Britain, against the Morini, who had revolted; who, as they had no place to which they might retreat, on account of the drying up of their marshes (which they had availed themselves of as a place of refuge the preceding year), almost all fell into the power of Labienus. In the meantime Caesar's lieutenants, Q. Titurius and L. Cotta, who had led the legions into the territories of the Menapii, having laid waste all their lands, cut down their corn and burnt their houses, returned to Caesar because the Menapii had all concealed themselves in their thickest woods. Caesar fixed the winter-quarters of all the legions amongst the Belgae. Thither only two British states sent hostages; the rest omitted to do so. For these successes, a thanksgiving of twenty days was decreed by the senate upon receiving Caesar's letter.

Book V

I. — Lucius Domitius and Appius Claudius being consuls, Caesar when departing from his winter-quarters into Italy, as he had been accustomed to do yearly, commands the lieutenants whom he appointed over the legions to take care that during the winter as many ships as possible should be built, and the old repaired. He plans the size and shape of them. For despatch of lading, and for drawing them on shore, he makes them a little lower than those which we have been accustomed to use in our sea; and that so much the more, because he knew that, on account of the frequent changes of the tide, less swells occurred there; for the purpose of transporting little and a great number of horses, [he makes them] a little broader than those which we use in other seas. All these he orders to be constructed for lightness and expedition, to which object their lowness contributes greatly. He orders those things which are necessary for equipping ships to be brought thither from Spain. He himself, on the assizes of Hither Gaul being concluded, proceeds into Illyricum, because he heard that the part of the province nearest them was being laid waste by the incursions of the Pirustae. When he had arrived there, he levies soldiers upon the states, and orders them to assemble at an appointed place. Which circumstance having been reported [to them], the Pirustae send ambassadors to him to inform him that no part of those proceedings was done by public deliberation, and assert that they were ready to make compensation by all means for the injuries [inflicted]. Caesar, accepting their defence, demands hostages, and orders them to be brought to him on a specified day, and assures them that unless they did so he would visit their state with war. These being brought to him on the day which he had ordered, he appoints arbitrators between the states, who should estimate the damages and determine the reparation.

II. — These things being finished, and the assizes being concluded, he returns into Hither Gaul, and proceeds thence to the army. When he had arrived there, having made a survey of the winter quarter, he finds that, by the extraordinary ardour of the soldiers, amidst the utmost scarcity of all materials, about six hundred ships of that kind which we have described above, and twenty-eight ships of war, had been built, and were not far from that state that they might be launched in a few days. Having commended the soldiers and those who had presided over the work, he informs them what he wishes to be done, and orders all the ships to assemble at port Itius, from which port he had learned that the passage into Britain was shortest, [being only] about thirty miles from the continent. He left what seemed a sufficient number of soldiers for that design; he himself proceeds into the territories of the Treviri with four legions without baggage, and 800 horse, because they neither came to the general diets [of Gaul], nor obeyed his commands, and were, moreover, said to be tampering with the Germans beyond the Rhine.

III. — This state is by far the most powerful of all Gaul in cavalry, and has great forces of infantry, and as we have remarked above, borders on the Rhine. In that state, two persons, Indutiomarus and Cingetorix, were then contending with each other for the supreme power; one of whom, as soon as the arrival of Caesar and his legions was known, came to him; assures him that he and all his party would continue in their allegiance, and not revolt from the alliance of the Roman people, and informs him of the things which were going on amongst the Treviri. But Indutiomarus began to collect cavalry and infantry, and make preparations for war, having concealed those who by reason of their age could not be under arms in the forest Arduenna, which is of immense size, [and] extends from the Rhine across the country of the Treviri to the frontiers of the Remi. But after that, some of the chief persons of the state, both influenced by their friendship for Cingetorix, and alarmed at the arrival of our army, came to Caesar and began to solicit him privately about their own interests, since they could not provide for the safety of the state; Indutiomarus, dreading lest he should be abandoned by all, sends ambassadors to Caesar, to declare that he absented himself from his countrymen, and refrained from coming to him on this account, that he might the more easily keep the state in its allegiance, lest on the departure of all the nobility the commonalty should, in their indiscretion, revolt. And thus the whole state was at his control; and that he, if Caesar would permit, would come to the camp to him, and would commit his own fortunes and those of the state to his good faith.

IV. — Caesar, though he discerned from what motive these things were said, and what circumstance deterred him from his meditated plan, still, in order that he might not be compelled to waste the summer among the Treviri, while all things were prepared for the war with Britain, ordered Indutiomarus to come to him with 200 hostages. When these were brought, [and] among them his son and near relations whom he had demanded by name, he consoled Indutiomarus, and enjoined him to continue in his allegiance; yet, nevertheless, summoning to him the chief men of the Treviri, he reconciled them individually to Cingetorix: this he both thought should be done by him in justice to the merits of the latter, and also judged that it was of great importance that the influence of one whose singular attachment towards him he had fully seen, should prevail as much as possible among his people. Indutiomarus was very much offended at this act, [seeing that] his influence was diminished among his countrymen; and he, who already before had borne a hostile mind towards us, was much more violently inflamed against us through resentment at this.

V. — These matters being settled, Caesar went to port Itius with the legions. There he discovers that forty ships which had been built in the country of the Meldi, having been driven back by a storm, had been unable to maintain their course, and had returned to the same port from which they had set out; he finds the rest ready for sailing, and furnished with everything. In the same place, the cavalry of the whole of Gaul, in number 4,000, assembles, and [also] the chief persons of all the states; he had determined to leave in Gaul a very few of them, whose fidelity towards him he had clearly discerned, and take the rest with him as hostages; because he feared a commotion in Gaul when he should be absent.

VI. — There was together with the others, Dumnorix, the Aeduan, of whom we have made previous mention. Him in particular he had resolved to have with him, because he had discovered him to be fond of change, fond of power, possessing great resolution, and great influence among the Gauls. To this was added that Dumnorix had before said in an assembly of Aeduans, that the sovereignty of the state had been made over to him by Caesar; which speech the Aedui bore with impatience and yet dared not send ambassadors to Caesar for the purpose of either rejecting or deprecating [that appointment]. That fact Caesar had learned from his own personal friends. He at first strove to obtain by every entreaty that he should be left in Gaul; partly, because, being unaccustomed to sailing, he feared the sea; partly, because he said he was prevented by divine admonitions. After he saw that this request was firmly refused him, all hope of success being lost, he began to tamper with the chief persons of the Gauls, to call them apart singly and exhort them to remain on the continent; to agitate them with the fear that it was not without reason that Gaul should be stript of all her nobility; that it was Caesar's design to bring over to Britain and put to death all those whom he feared to slay in the sight of Gaul, to pledge his honour to the rest, to ask for their oath that they would by common deliberation execute what they should perceive to be necessary for Gaul. These things were reported to Caesar by several persons.

VII. — Having learned this fact, Caesar, because he had conferred so much honour upon the Aeduan state, determined that Dumnorix should be restrained and deterred by whatever means he could; and that, because he perceived his insane designs to be proceeding farther and farther, care should be taken lest he might be able to injure him and the commonwealth. Therefore, having stayed about twenty-five days in that place, because the north wind, which usually blows a great part of every season, prevented the voyage, he exerted himself to keep Dumnorix in his allegiance [and] nevertheless learn all his measures: having at length met with favourable weather, he orders the foot soldiers and the horse to embark in the ships. But, while the minds of all were occupied, Dumnorix began to take his departure from the camp homewards with the cavalry of the Aedui, Caesar being ignorant of it. Caesar, on this matter being reported to him, ceasing from his expedition and deferring all other affairs, sends a great part of the cavalry to pursue him, and commands that he be brought back; he orders that if he uses violence and does not submit, that he be slain: considering that Dumnorix would do nothing as a rational man while he himself was absent, since he had disregarded his command even when present. He, however, when recalled, began to resist and defend himself with his hand, and implore the support of his people, often exclaiming that "he was free and the subject of a free state." They surround and kill the man as they had been commanded; but the Aeduan horsemen all return to Caesar.

VIII. — When these things were done [and] Labienus, left on the continent with three legions and 2,000 horse, to defend the harbours and provide corn, and discover what was going on in Gaul, and take measures according to the occasion and according to the circumstance; he himself, with five legions and a number of horse, equal to that which he was leaving on the continent, set sail at sunset and [though for a time] borne forward by a gentle south-west wind, he did not maintain his course, in consequence of the wind dying away about midnight, and being carried on too far by the tide, when the sun rose, espied Britain passed on his left. Then, again, following the change of tide, he urged on with the oars that he might make that port of the island in which he had discovered the preceding summer that there was the best landing-place, and in this affair the spirit of our soldiers was very much to be extolled; for they with the transports and heavy ships, the labour of rowing not being [for a moment] discontinued, equalled the speed of the ships of war. All the ships reached Britain nearly at mid-day; nor was there seen a [single] enemy in that place, but, as Caesar afterwards found from some prisoners, though large bodies of troops had assembled there, yet being alarmed by the great number of our ships, more than eight hundred of which, including the ships of the preceding year, and those private vessels which each had built for his own convenience, had appeared at one time, they had quitted the coast and concealed themselves among the higher points.

IX. — Caesar, having disembarked his army and chosen a convenient place for the camp, when he discovered from the prisoners in what part the forces of the enemy had lodged themselves, having left ten cohorts and 300 horse at the sea, to be a guard to the ships, hastens to the enemy, at the third watch, fearing the less for the ships for this reason, because he was leaving them fastened at anchor upon an even and open shore; and he placed Q. Atrius over the guard of the ships. He himself, having advanced by night about twelve miles, espied the forces of the enemy. They, advancing to the river with their cavalry and chariots from the higher ground, began to annoy our men and give battle. Being repulsed by our cavalry, they concealed themselves in woods, as they had secured a place admirably fortified by nature and by art, which, as it seemed, they had before prepared on account of a civil war; for all entrances to it were shut up by a great number of felled trees. They themselves rushed out of the woods to fight here and there, and prevented our men from entering their fortifications. But the soldiers of the seventh legion, having formed a testudo and thrown up a rampart against the fortification, took the place and drove them out of the woods, receiving only a few wounds. But Caesar forbade his men to pursue them in their flight any great distance; both because he was ignorant of the nature of the ground, and because, as a great part of the day was spent, he wished time to be left for the fortification of the camp.

X. — The next day, early in the morning, he sent both foot-soldiers and horse in three divisions on an expedition to pursue those who had fled. These having advanced a little way, when already the rear [of the enemy] was in sight, some horse came to Caesar from Quintus Atrius, to report that the preceding night, a very great storm having arisen, almost all the ships were dashed to pieces and cast upon the shore, because neither the anchors and cables could resist, nor could the sailors and pilots sustain the violence of the storm; and thus great damage was received by that collision of the ships.

XI. — These things being known [to him], Caesar orders the legions and cavalry to be recalled and to cease from their march; he himself returns to the ships: he sees clearly before him almost the same things which he had heard of from the messengers and by letter, so that, about forty ships being lost, the remainder seemed capable of being repaired with much labour. Therefore he selects workmen from the legions, and orders others to be sent for from the continent; he writes to Labienus to build as many ships as he could with those legions which were with him. He himself, though the matter was one of great difficulty and labour, yet thought it to be most expedient for all the ships to be brought up on shore and joined with the camp by one fortification. In these matters he employed about ten days, the labour of the soldiers being unremitting even during the hours of night. The ships having been brought up on shore and the camp strongly fortified, he left the same forces which he did before as a guard for the ships; he sets out in person for the same place that he had returned from. When he had come thither, greater forces of the Britons had already assembled at that place, the chief command and management of the war having been entrusted to Cassivellaunus, whose territories a river, which is called the Thames, separates from the maritime states at about eighty miles from the sea. At an earlier period perpetual wars had taken place between him and the other states; but, greatly alarmed by our arrival, the Britons had placed him over the whole war and the conduct of it.

XII. — The interior portion of Britain is inhabited by those of whom they say that it is handed down by tradition that they were born in the island itself: the maritime portion by those who had passed over from the country of the Belgae for the purpose of plunder and making war; almost all of whom are called by the names of those states from which being sprung they went thither, and having waged war, continued there and began to cultivate the lands. The number of the people is countless, and their buildings exceedingly numerous, for the most part very like those of the Gauls: the number of cattle is great. They use either brass or iron rings, determined at a certain weight, as their money. Tin is produced in the midland regions; in the maritime, iron; but the quantity of it is small: they employ brass, which is imported. There, as in Gaul, is timber of every description, except beech and fir. They do not regard it lawful to eat the hare, and the cock, and the goose; they, however, breed them for amusement and pleasure. The climate is more temperate than in Gaul, the colds being less severe.

XIII. — The island is triangular in its form, and one of its sides is opposite to Gaul. One angle of this side, which is in Kent, whither almost all ships from Gaul are directed, [looks] to the east; the lower looks to the south. This side extends about 500 miles. Another side lies towards Spain and the west, on which part is Ireland, less, as is reckoned, than Britain by one-half; but the passage [from it] into Britain is of equal distance with that from Gaul. In the middle of this voyage is an island, which is called Mona; many smaller islands besides are supposed to lie [there], of which islands some have written that at the time of the winter solstice it is night there for thirty consecutive days. We, in our inquiries about that matter, ascertained nothing, except that, by accurate measurements with water, we perceived the nights to be shorter there than on the continent. The length of this side, as their account states, is 700 miles. The third side is towards the north, to which portion of the island no land is opposite; but an angle of that side looks principally towards Germany. This side is considered to be 800 miles in length. Thus the whole island is [about] 2000 miles in circumference.

XIV. — The most civilised of all these nations are they who inhabit Kent, which is entirely a maritime district, nor do they differ much from the Gallic customs. Most of the inland inhabitants do not sow corn, but live on milk and flesh, and are clad with skins. All the Britons, indeed, dye themselves with wood, which occasions a bluish colour, and thereby have a more terrible appearance in fight. They wear their hair long, and have every part of their body shaved except their head and upper lip. Ten and even twelve have wives common to them, and particularly brothers among brothers, and parents among their children; but if there be any issue by these wives, they are reputed to be the children of those by whom respectively each was first espoused when a virgin.

XV. — The horse and charioteers of the enemy contended vigorously in a skirmish with our cavalry on the march; yet so that our men were conquerors in all parts, and drove them to their woods and hills; but, having slain a great many, they pursued too eagerly, and lost some of their men. But the enemy, after some time had elapsed, when our men were off their guard, and occupied in the fortification of the camp, rushed out of the woods, and making an attack upon those who were placed on duty before the camp, fought in a determined manner; and two cohorts being sent by Caesar to their relief, and these severally the first of two legions, when these had taken up their position at a very small distance from each other, as our men were disconcerted by the unusual mode of battle, the enemy broke through the middle of them most courageously, and retreated thence in safety. That day, Q. Laberius Durus, a tribune of the soldiers, was slain. The enemy, since more cohorts were sent against them, were repulsed.

XVI. — In the whole of this method of fighting since the engagement took place under the eyes of all and before the camp, it was perceived that our men, on account of the weight of their arms, inasmuch as they could neither pursue [the enemy when] retreating, nor dare quit their standards, were little suited to this kind of enemy; that the horse also fought with great danger, because they [the Britons] generally retreated even designedly, and, when they had drawn off our men a short distance from the legions, leaped from their chariots and fought on foot in unequal [and to them advantageous] battle. But the system of cavalry engagement is wont to produce equal danger, and indeed the same, both to those who retreat and those who pursue. To this was added, that they never fought in close order, but in small parties and at great distances, and had detachments placed [in different parts], and then the one relieved the other, and the vigorous and fresh succeeded the wearied.

XVII. — The following day the enemy halted on the hills, a distance from our camp, and presented themselves in small parties, and began to challenge our horse to battle with less spirit than the day before. But at noon, when Caesar had sent three legions, and all the cavalry with C. Trebonius, the lieutenant, for the purpose of foraging, they flew upon the foragers suddenly from all quarters, so that they did not keep off [even] from the standards and the legions. Our men making an attack on them vigorously, repulsed them; nor did they cease to pursue them until the horse, relying on relief, as they saw the legions behind them, drove the enemy precipitately before them, and, slaying a great number of them, did not give them the opportunity either of rallying or halting, or leaping from their chariots. Immediately after this retreat, the auxiliaries who had assembled from all sides, departed; nor after that time did the enemy ever engage with us in very large numbers.

XVIII. — Caesar, discovering their design, leads his army into the territories of Cassivellaunus to the river Thames; which river can be forded in one place only, and that with difficulty. When he had arrived there, he perceives that numerous forces of the enemy were marshalled on the other bank of the river; the bank also was defended by sharp stakes fixed in front, and stakes of the same kind fixed under the water were covered by the river. These things being discovered from [some] prisoners and deserters, Caesar, sending forward the cavalry, ordered the legions to follow them immediately. But the soldiers advanced with such speed and such ardour, though they stood above the water by their heads only, that the enemy could not sustain the attack of the legions and of the horse, and quitted the banks, and committed themselves to flight.

XIX. — Cassivellaunus, as we have stated above, all hope [rising out] of battle being laid aside, the greater part of his forces being dismissed, and about 4,000 charioteers only being left, used to observe our marches and retire a little from the road, and conceal himself in intricate and woody places, and in those neighbourhoods in which he had discovered we were about to march, he used to drive the cattle and the inhabitants from the fields into the woods; and, when our cavalry, for the sake of plundering and ravaging the more freely, scattered themselves among the fields, he used to send out charioteers from the woods by all the well-known roads and paths, and, to the great danger of our horse, engage with them; and this source of fear hindered them from straggling very extensively. The result was that Caesar did not allow excursions to be made to a great distance from the main body of the legions, and ordered that damage should be done to the enemy in ravaging their lands and kindling fires only so far as the legionary soldiers could, by their own exertion and marching, accomplish it.

XX. — In the meantime, the Trinobantes, almost the most powerful state of those parts, from which the young man Mandubratius embracing the protection of Caesar had come to the continent of Gaul to [meet] him (whose father, Imanuentius, had possessed the sovereignty in that state, and had been killed by Cassivellaunus; he himself had escaped death by flight), send ambassadors to Caesar, and promise that they will surrender themselves to him and perform his commands; they entreat him to protect Mandubratius from the violence of Cassivellaunus, and send to their state some one to preside over it, and possess the government. Caesar demands forty hostages from them, and corn for his army, and sends Mandubratius to them. They speedily performed the things demanded, and sent hostages to the number appointed, and the corn.

XXI. — The Trinobantes being protected and secured from any violence of the soldiers, the Cenimagni, the Segontiaci, the Ancalites, the Bibroci, and the Cassi, sending embassies, surrender themselves to Caesar.

From them he learns that the capital town of Cassivellaunus was not far from that place, and was defended by woods and morasses, and a very large number of men and of cattle had been collected in it. (Now the Britons, when they have fortified the intricate woods, in which they are wont to assemble for the purpose of avoiding the incursion of an enemy, with an entrenchment and a rampart, call them a town.) Thither he proceeds with his legions: he finds the place admirably fortified by nature and art; he, however, undertakes to attack it in two directions. The enemy, having remained only a short time, did not sustain the attack of our soldiers, and hurried away on the other side of the town. A great amount of cattle was found there, and many of the enemy were taken and slain in their flight.

XXII. — While these things are going forward in those places, Cassivellaunus sends messengers into Kent, which, we have observed above, is on the sea, over which districts four several kings reigned, Cingetorix, Carvilius, Taximagulus, and Segonax, and commands them to collect all their forces, and unexpectedly assail and storm the naval camp. When they had come to the camp, our men, after making a sally, slaying many of their men, and also capturing a distinguished leader named Lugotorix, brought back their own men in safety. Cassivellaunus, when this battle was reported to him, as so many losses had been sustained, and his territories laid waste, being alarmed most of all by the desertion of the states, sends ambassadors to Caesar [to treat] about a surrender through the mediation of Commius the Atrebatian. Caesar, since he had determined to pass the winter on the continent, on account of the sudden revolts of Gaul, and as much of the summer did not remain, and he perceived that even that could be easily protracted, demands hostages, and prescribes what tribute Britain should pay each year to the Roman people; he forbids and commands Cassivellaunus that he wage not war against Mandubratius or the Trinobantes.

XXIII. — When he had received the hostages, he leads back the army to the sea, and finds the ships repaired. After launching these, because he had a large number of prisoners, and some of the ships had been lost in the storm, he determines to convey back his army at two embarkations. And it so happened, that out of so large a number of ships, in so many voyages, neither in this nor in the previous year was any ship missing which conveyed soldiers; but very few out of those which were sent back to him from the continent empty, as the soldiers of the former convoy had been disembarked, and out of those (sixty in number) which Labienus had taken care to have built, reached their destination; almost all the rest were driven back, and when Caesar had waited for them for some time in vain, lest he should be debarred from a voyage by the season of the year, inasmuch as the equinox was at hand, he of necessity stowed his soldiers the more closely, and, a very great calm coming on, after he had weighed anchor at the beginning of the second watch, he reached land at break of day and brought in all the ships in safety.

XXIV. — The ships having been drawn up and a general assembly of the Gauls held at Samarobriva, because the corn that year had not prospered in Gaul by reason of the droughts, he was compelled to station his army in its winter-quarters, differently from the former years, and to distribute the legions among several states: one of them he gave to C. Fabius, his lieutenant, to be marched into the territories of the Morini; a second to Q. Cicero, into those of the Nervii; a third to L. Roscius, into those of the Essui; a fourth he ordered to winter with T. Labienus among the Remi in the confines of the Treviri; he stationed three in Belgium; over these he appointed M. Crassus, his questor, and L. Munatius Plancus and C. Trebonius, his lieutenants. One legion which he had raised last on the other side of the Po, and five cohorts, he sent amongst the Eburones, the greatest portion of whom lie between the Meuse and the Rhine, [and] who were under the government of Ambiorix and Cativolcus. He ordered Q. Titurius Sabinus and L. Aurunculeius Cotta, his lieutenants, to take the command of these soldiers. The legions being distributed in this manner, he thought he could most easily remedy the scarcity of corn; and yet the winter-quarters of all these legions (except that which he had given to L. Roscius to be led into the most peaceful and tranquil neighbourhood) were comprehended within [about] 100 miles. He himself in the meanwhile, until he had stationed the legions and knew that the several winter-quarters were fortified, determined to stay in Gaul.

XXV. — There was among the Carnutes a man named Tasgetius, born of very high rank, whose ancestors had held the sovereignty in his state. To him Caesar had restored the position of his ancestors, in consideration of his prowess and attachment towards him, because in all his wars he had availed himself of his valuable services. His personal enemies had killed him when in the third year of his reign, many even of his own state being openly promoters [of that act]. This event is related to Caesar. He fearing, because several were involved in the act, that the state might revolt at their instigation, orders Lucius Plancus, with a legion, to proceed quickly from Belgium to the Carnutes, and winter there, and arrest and send to him the persons by whose instrumentality he should discover that Tasgetius was slain. In the meantime, he was apprised by all the lieutenants and questors to whom he had assigned the legions, that they had arrived in winter-quarters, and that the place for the quarters was fortified.

XXVI. — About fifteen days after they had come into winter-quarters, the beginning of a sudden insurrection and revolt arose from Ambiorix and Cativolcus, who, though they had met with Sabinus and Cotta at the borders of their kingdom, and had conveyed corn into our winter-quarters, induced by the messages of Indutiomarus, one of the Treviri, excited their people, and after having suddenly assailed the soldiers, engaged in procuring wood, came with a large body to attack the camp.

When our men had speedily taken up arms and had ascended the rampart, and sending out some Spanish horse on one side, had proved conquerors in a cavalry action, the enemy, despairing of success, drew off their troops from the assault. Then they shouted, according to their custom, that some of our men should go forward to a conference, [alleging] that they had some things which they desired to say respecting the common interest, by which they trusted their disputes could be removed.

XXVII. — C. Arpineius, a Roman knight, the intimate friend of Q. Titurius, and with him Q. Junius, a certain person from Spain, who already on previous occasions had been accustomed to go to Ambiorix, at Caesar's mission, is sent to them for the purpose of a conference: before them Ambiorix spoke to this effect: "That he confessed that for Caesar's kindness towards him he was very much indebted to him, inasmuch as by his aid he had been freed from a tribute which he had been accustomed to pay to the Aduatuci, his neighbours; and because his own son and the son of his brother had been sent back to him, whom, when sent in the number of hostages, the Aduatuci had detained among them in slavery and in chains; and that he had not done that which he had done in regard to the attacking of the camp, either by his own judgment or desire, but by the compulsion of his state; and that his government was of that nature, that the people had as much of authority over him as he over the people. To the state moreover the occasion of the war was this — that it could not withstand the sudden combination of the Gauls; that he could easily prove this from his own weakness, since he was not so little versed in affairs as to presume that with his forces he could conquer the Roman people; but that it was the common resolution of Gaul; that that day was appointed for the storming of all Caesar's winter-quarters, in order that no legion should be able to come to the relief of another legion, that Gauls could not easily deny Gauls, especially when a measure seemed entered into for recovering their common freedom. Since he had performed his duty to them on the score of patriotism [he said], he has now regard to gratitude for the kindness of Caesar; that he warned, that he prayed Titurius by the claims of hospitality, to consult for his and his soldiers' safety; that a large force of the Germans had been hired and had passed the Rhine; that it would arrive in two days; that it was for them to consider whether they thought fit, before the nearest people perceived it, to lead off their soldiers when drawn out of winter-quarters, either to Cicero or to Labienus; one of whom was about fifty miles distant from them, the other rather more; that this he promised and confirmed by oath, that he would give them a safe passage through his territories; and when he did that, he was both consulting for his own state, because it would be relieved from the winter-quarters, and also making a requital to Caesar for his obligations."

XXVIII. — Arpineius and Junius relate to the lieutenants what they had heard. They, greatly alarmed by the unexpected affair, though those things were spoken by an enemy, still thought they were not to be disregarded; and they were especially influenced by this consideration, that it was scarcely credible that the obscure and humble state of the Eburones had dared to make war upon the Roman people of their own accord. Accordingly, they refer the matter to a council, and a, great controversy arises among them. L. Aurunculeius, and several tribunes of the soldiers and the centurions of the first rank, were of opinion "that nothing should be done hastily, and that they should not depart from the camp without Caesar's orders"; they declared, "that any forces of the Germans, however great, might be encountered by fortified winter-quarters; that this fact was a proof [of it]; that they had sustained the first assault of the Germans most valiantly, inflicting many wounds upon them; that they were not distressed for corn; that in the meantime relief would come both from the nearest winter-quarters and from Caesar"; lastly, they put the query, "what could be more undetermined, more undignified, than to adopt measures respecting the most important affairs on the authority of an enemy?"

XXIX. — In opposition to those things Titurius exclaimed, "That they would do this too late, when greater forces of the enemy, after a junction with the Germans, should have assembled; or when some disaster had been received in the neighbouring winter-quarters; that the opportunity for deliberating was short; that he believed that Caesar had set forth into Italy, as the Carnutes would not otherwise have taken the measure of slaying Tasgetius, nor would the Eburones, if he had been present, have come to the camp with so great defiance of us; that he did not regard the enemy, but the fact, as the authority; that the Rhine was near; that the death of Ariovistus and our previous victories were subjects of great indignation to the Germans; that Gaul was inflamed, that after having received so many defeats she was reduced under the sway of the Roman people, her pristine glory in military matters being extinguished." Lastly, "who would persuade himself of this, that Ambiorix had resorted to a design of that nature without sure grounds? That his own opinion was safe on either side; if there be nothing very formidable, they would go without danger to the nearest legion; if all Gaul conspired with the Germans, their only safety lay in despatch. What issue would the advice of Cotta and of those who differed from him, have? from which, if immediate danger was not to be dreaded, yet certainly famine, by a protracted siege, was."

XXX. — This discussion having been held on the two sides, when opposition was offered strenuously by Cotta and the principal officers, "Prevail," said Sabinus, "if so you wish it"; and he said it with a louder voice, that a great portion of the soldiers might hear him; "nor am I the person among you," he said, "who is most powerfully alarmed by the danger of death; these will be aware of it, and then, if any thing disastrous shall have occurred, they will demand a reckoning at your hands; these, who, if it were permitted by you, united three days hence with the nearest winter-quarters, may encounter the common condition of war with the rest, and not, as if forced away and separated far from the rest, perish either by the sword or by famine."

XXXI. — They rise from the council, detain both, and entreat, that "they do not bring the matter into the greatest jeopardy by their dissension and obstinacy; the affair was an easy one, if only they all thought and approved of the same thing, whether they remain or depart; on the other hand, they saw no security in dissension." The matter is prolonged by debate till midnight. At last Cotta, being overruled, yields his assent; the opinion of Sabinus prevails. It is proclaimed that they will march at day-break; the remainder of the night is spent without sleep, since every soldier was inspecting his property, [to see] what he could carry with him, and what, out of the appurtenances of the winter-quarters, he would be compelled to leave; every reason is suggested to show why they could not stay without danger, and how that danger would be increased by the fatigue of the soldiers and their want of sleep. At break of day they quit the camp, in a very extended line and with a very large amount of baggage, in such a manner as men who were convinced that the advice was given by Ambiorix, not as an enemy, but as most friendly [towards them].

XXXII. — But the enemy, after they had made the discovery of their intended departure by the noise during the night and their not retiring to rest, having placed an ambuscade in two divisions in the woods, in a suitable and concealed place, two miles from the camp, waited for the arrival of the Romans; and when the greater part of the line of march had descended into a considerable valley, they suddenly presented themselves on either side of that valley, and began both to harass the rear and hinder the van from ascending, and to give battle in a place exceedingly disadvantageous to our men.

XXXIII. — Then at length Titurius, as one who had provided nothing beforehand, was confused, ran to and fro, and set about arranging his troops; these very things, however, he did timidly and in such a manner that all resources seemed to fail him: which generally happens to those who are compelled to take council in the action itself. But Cotta, who had reflected that these things might occur on the march, and on that account had not been an adviser of the departure, was wanting to the common safety in no respect; both in addressing and encouraging the soldiers, he performed the duties of a general, and in the battle those of a soldier. And since they [Titurius and Cotta] could less easily perform everything by themselves, and provide what was to be done in each place, by reason of the length of the line of march, they ordered [the officers] to give the command that they should leave the baggage and form themselves into an orb, which measure, though in a contingency of that nature it was not to be condemned, still turned out unfortunately; for it both diminished the hope of our soldiers and rendered the enemy more eager for the fight, because it appeared that this was not done without the greatest fear and despair. Besides that happened, which would necessarily be the case, that the soldiers for the most part quitted their ensigns and hurried to seek and carry off from the baggage whatever each thought valuable, and all parts were filled with uproar and lamentation.

XXXIV. — But judgment was not wanting to the barbarians; for their leaders ordered [the officers] to proclaim through the ranks "that no man should quit his place; that the booty was theirs, and for them was reserved whatever the Romans should leave; therefore let them consider that all things depended on their victory." Our men were equal to them in fighting, both in courage and in number, and though they were deserted by their leader and by fortune, yet they still placed all hope of safety in their valour, and as often as any cohort sallied forth on that side, a great number of the enemy usually fell. Ambiorix, when he observed this, orders the command to be issued that they throw their weapons from a distance and do not approach too near, and in whatever direction the Romans should make an attack, there give way (from the lightness of their appointments and from their daily practice no damage could be done them); [but] pursue them when betaking themselves to their standards again.

XXXV. — Which command having been most carefully obeyed, when any cohort had quitted the circle and made a charge, the enemy fled very precipitately. In the meantime, that part of the Roman army, of necessity, was left unprotected, and the weapons received on their open flank. Again, when they had begun to return to that place from which they had advanced, they were surrounded both by those who had retreated and by those who stood next them; but if, on the other hand, they wished to keep their place, neither was an opportunity left for valour, nor could they, being crowded together, escape the weapons cast by so large a body of men. Yet, though assailed by so many disadvantages, [and] having received many wounds, they withstood the enemy, and, a great portion of the day being spent, though they fought from day-break till the eighth hour, they did nothing which was unworthy of them. At length, each thigh of T. Balventius, who the year before had been chief centurion, a brave man and one of great authority, is pierced with a javelin; Q. Lucanius, of the same rank, fighting most valiantly, is slain while he assists his son when surrounded by the enemy; L. Cotta, the lieutenant, when encouraging all the cohorts and companies, is wounded full in the mouth by a sling.

XXXVI. — Much troubled by these events, Q. Titurius, when he had perceived Ambiorix in the distance encouraging his men, sends to him his interpreter, Cn. Pompey, to beg that he would spare him and his soldiers.

He, when addressed, replied, "If he wished to confer with him, it was permitted; that he hoped what pertained to the safety of the soldiers could be obtained from the people; that to him however certainly no injury would be done, and that he pledged his faith to that effect." He consults with Cotta, who had been wounded, whether it would appear right to retire from battle, and confer with Ambiorix; [saying] that he hoped to be able to succeed respecting his own and the soldiers' safety. Cotta says he will not go to an armed enemy, and in that perseveres.

XXXVII. — Sabinus orders those tribunes of the soldiers whom he had at the time around him, and the centurions of the first ranks, to follow him, and when he had approached near to Ambiorix, being ordered to throw down his arms, he obeys the order and commands his men to do the same. In the meantime, while they treat upon the terms, and a longer debate than necessary is designedly entered into by Ambiorix, being surrounded by degrees, he is slain. Then they according to their custom shout out "Victory," and raise their war-cry, and, making an attack on our men, break their ranks. There L. Cotta, while fighting, is slain, together with the greater part of the soldiers; the rest betake themselves to the camp from which they had marched forth, and one of them, L. Petrosidius, the standard bearer, when he was overpowered by the great number of the enemy, threw the eagle within the entrenchments and is himself slain while fighting with the greatest courage before the camp. They with difficulty sustain the attack till night; despairing of safety, they all to a man destroy themselves in the night. A few escaping from the battle, make their way to Labienus at winter-quarters, after wandering at random through the woods, and inform him of these events.

XXXVIII. — Elated by this victory, Ambiorix marches immediately with his cavalry to the Aduatuci, who bordered on his kingdom; he halts neither day nor night, and orders the infantry to follow him closely. Having related the exploit and roused the Aduatuci, the next day he arrived among the Nervii, and entreats "that they should not throw away the opportunity of liberating themselves for ever and of punishing the Romans for those wrongs which they had received from them"; [he tells them] "that two lieutenants have been slain, and that a large portion of the army has perished; that it was not a matter of difficulty for the legion which was wintering with Cicero to be cut off, when suddenly assaulted; he declares himself ready to co-operate in that design." He easily gains over the Nervii by this speech.

XXXIX. — Accordingly, messengers having been forthwith despatched to the Centrones, the Grudii, the Levaci, the Pleumoxii, and the Geiduni, all of whom are under their government, they assemble as large bodies as they can, and rush unexpectedly to the winter-quarters of Cicero, the report of the death of Titurius not having as yet been conveyed to him. That also occurred to him which was the consequence of a necessary work, — that some soldiers who had gone off into the woods for the purpose of procuring timber and therewith constructing fortifications, were intercepted by the sudden arrival of [the enemy's] horse. These having been entrapped, the Eburones, the Nervii, and the Aduatuci and all their allies and dependants, begin to attack the legion: our men quickly run together to arms and mount the rampart: they sustained the attack that day with great difficulty, since the enemy placed all their hope in despatch, and felt assured that, if they obtained this victory, they would be conquerors for ever.

XL. — Letters are immediately sent to Caesar by Cicero, great rewards being offered [to the messengers] if they carried them through. All the passes having been beset, those who were sent are intercepted. During the night as many as 120 towers are raised with incredible despatch out of the timber which they had collected for the purpose of fortification: the things which seemed necessary to the work are completed. The following day the enemy, having collected far greater forces, attack the camp [and] fill up the ditch. Resistance is made by our men in the same manner as the day before: this same thing is done afterwards during the remaining days. The work is carried on incessantly in the night: not even to the sick, or wounded, is opportunity given for rest: whatever things are required for resisting the assault of the next day are provided during the night: many stakes burnt at the end, and a large number of mural pikes are procured: towers are built up, battlements and parapets are formed of interwoven hurdles. Cicero himself, though he was in very weak health, did not leave himself the night-time for repose, so that he was forced to spare himself by the spontaneous movement and entreaties of the soldiers.

XLI. — Then these leaders and chiefs of the Nervii, who had any intimacy and grounds of friendship with Cicero, say they desire to confer with him. When permission was granted, they recount the same things which Ambiorix had related to Titurius, namely, "that all Gaul was in arms, that the Germans had passed the Rhine, that the winter-quarters of Caesar and of the others were attacked." They report in addition also, about the death of Sabinus. They point to Ambiorix for the purpose of obtaining credence; "they are mistaken," say they, "if they hoped for any relief from those who distrust their own affairs; that they bear such feelings towards Cicero and the Roman people that they deny them nothing but winter-quarters and are unwilling that this practice should become constant; that through their [the Nervii's] means it is possible for them [the Romans] to depart from their winter-quarters safely and to proceed without fear into whatever parts they desire." To these Cicero made only one reply: "that it is not the custom of the Roman people to accept any condition from an armed enemy: if they are willing to lay down their arms, they may employ him as their advocate and send ambassadors to Caesar: that he believed, from his [Caesar's] justice, they would obtain the things which they might request."

XLII. — Disappointed in this hope, the Nervii surround the winter-quarters with a rampart eleven feet high, and a ditch thirteen feet in depth. These military works they had learnt from our men in the intercourse of former years, and, having taken some of our army prisoners, were instructed by them: but, as they had no supply of iron tools which are requisite for this service, they were forced to cut the turf with their swords, and to empty out the earth with their hands and cloaks, from which circumstance the vast number of the men could be inferred; for in less than three hours they completed a fortification of ten miles in circumference; and during the rest of the days they began to prepare and construct towers of the height of the ramparts, and grappling irons, and mantlets, which the same prisoners had taught them.

XLIII. — On the seventh day of the attack, a very high wind having sprung up, they began to discharge by their slings hot balls made of burnt or hardened clay, and heated javelins, upon the huts, which, after the Gallic custom, were thatched with straw. These quickly took fire, and by the violence of the wind, scattered their flames in every part of the camp. The enemy following up their success with a very loud shout, as if victory were already obtained and secured, began to advance their towers and mantlets, and climb the rampart with ladders. But so great was the courage of our soldiers, and such their presence of mind, that though they were scorched on all sides, and harassed by a vast number of weapons, and were aware that their baggage and their possessions were burning, not only did no one quit the rampart for the purpose of withdrawing from the scene, but scarcely did any one even then look behind; and they all fought most vigorously and most valiantly. This day was by far the most calamitous to our men; it had this result, however, that on that day the largest number of the enemy was wounded and slain, since they had crowded beneath the very rampart, and the hindmost did not afford the foremost a retreat. The flame having abated a little, and a tower having been brought up in a particular place and touching the rampart, the centurions of the third cohort retired from the place in which they were standing, and drew off all their men: they began to call on the enemy by gestures and by words, to enter if they wished; but none of them dared to advance. Then stones having been cast from every quarter, the enemy were dislodged, and their tower set on fire.

XLIV. — In that legion there were two very brave men, centurions, who were now approaching the first ranks, T. Pulfio, and L. Varenus. These used to have continual disputes between them which of them should be preferred, and every year used to contend for promotion with the utmost animosity. When the fight was going on most vigorously before the fortifications, Pulfio, one of them, says, "Why do you hesitate, Varenus? or what [better] opportunity of signalising your valour do you seek? This very day shall decide our disputes." When he had uttered these words, he proceeds beyond the fortifications, and rushes on that part of the enemy which appeared the thickest. Nor does Varenus remain within the rampart, but respecting the high opinion of all, follows close after. Then, when an inconsiderable space intervened, Pulfio throws his javelin at the enemy, and pierces one of the multitude who was running up, and while the latter was wounded and slain, the enemy cover him with their shields, and all throw their weapons at the other and afford him no opportunity of retreating. The shield of Pulfio is pierced and a javelin is fastened in his belt. This circumstance turns aside his scabbard and obstructs his right hand when attempting to draw his sword: the enemy crowd around him when [thus] embarrassed. His rival runs up to him and succours him in this emergency. Immediately the whole host turn from Pulfio to him, supposing the other to be pierced through by the javelin. Varenus rushes on briskly with his sword and carries on the combat hand to hand, and having slain one man, for a short time drove back the rest: while he urges on too eagerly, slipping into a hollow, he fell. To him, in his turn, when surrounded, Pulfio brings relief; and both having slain a great number, retreat into the fortifications amidst the highest applause. Fortune so dealt with both in this rivalry and conflict, that the one competitor was a succour and a safeguard to the other, nor could it be determined which of the two appeared worthy of being preferred to the other.

XLV. — In proportion as the attack became daily more formidable and violent, and particularly because, as a great number of the soldiers were exhausted with wounds, the matter had come to a small number of defenders, more frequent letters and messengers were sent to Caesar; a part of which messengers were taken and tortured to death in the sight of our soldiers. There was within our camp a certain Nervian, by name Vertico, born in a distinguished position, who in the beginning of the blockade had deserted to Cicero, and had exhibited his fidelity to him. He persuades his slave, by the hope of freedom, and by great rewards, to convey a letter to Caesar. This he carries out bound about his javelin, and mixing among the Gauls without any suspicion by being a Gaul, he reaches Caesar. From him they received information of the imminent danger of Cicero and the legion.

XLVI. — Caesar having received the letter about the eleventh hour of the day, immediately sends a messenger to the Bellovaci, to M. Crassus, questor there, whose winter-quarters were twenty-five miles distant from him. He orders the legion to set forward in the middle of the night and come to him with despatch. Crassus set out with the messenger. He sends another to C. Fabius, the lieutenant, ordering him to lead forth his legion into the territories of the Atrebates, to which he knew his march must be made. He writes to Labienus to come with his legion to the frontiers of the Nervii, if he could do so to the advantage of the commonwealth: he does not consider that the remaining portion of the army, because it was somewhat farther distant, should be waited for; but assembles about 400 horse from the nearest winter-quarters.

XLVII. — Having been apprised of the arrival of Crassus by the scouts at about the third hour, he advances twenty miles that day. He appoints Crassus over Samarobriva and assigns him a legion, because he was leaving there the baggage of the army, the hostages of the states, the public documents, and all the corn, which he had conveyed thither for passing the winter. Fabius, without delaying a moment, meets him on the march with his legion, as he had been commanded. Labienus, having learnt the death of Sabinus and the destruction of the cohorts, as all the forces of the Treviri had come against him, beginning to fear lest, if he made a departure from his winter-quarters, resembling a flight, he should not be able to support the attack of the enemy, particularly since he knew them to be elated by their recent victory, sends back a letter to Caesar, informing him with what great hazard he would lead out his legion from winter-quarters; he relates at large the affair which had taken place among the Eburones; he informs him that all the infantry and cavalry of the Treviri had encamped at a distance of only three miles from his own camp.

XLVIII. — Caesar, approving of his motives, although he was disappointed in his expectation of three legions, and reduced to two, yet placed his only hopes of the common safety in despatch. He goes into the territories of the Nervii by long marches. There he learns from some prisoners what things are going on in the camp of Cicero, and in how great jeopardy the affair is. Then with great rewards he induces a certain man of the Gallic horse to convey a letter to Cicero. This he sends written in Greek characters, lest the letter being intercepted, our measures should be discovered by the enemy. He directs him, if he should be unable to enter, to throw his spear with the letter fastened to the thong inside the fortifications of the camp. He writes in the letter, that he having set out with his legions, will quickly be there: he entreats him to maintain his ancient valour. The Gaul apprehending danger, throws his spear as he had been directed. It by chance stuck in a tower, and, not being observed by our men for two days, was seen by a certain soldier on the third day: when taken down, it was carried to Cicero. He, after perusing it, reads it out in an assembly of the soldiers, and fills all with the greatest joy. Then the smoke of the fires was seen in the distance, a circumstance which banished all doubt of the arrival of the legions.

XLIX. — The Gauls, having discovered the matter through their scouts, abandon the blockade, and march towards Caesar with all their forces: these were about 60,000 armed men. Cicero, an opportunity being now afforded, again begs of that Vertico, the Gaul, whom we mentioned above, to convey back a letter to Caesar; he advises him to perform his journey warily; he writes in the letter that the enemy had departed and had turned their entire force against him. When this letter was brought to him about the middle of the night, Caesar apprises his soldiers of its contents, and inspires them with courage for fighting: the following day, at the dawn, he moves his camp, and, having proceeded four miles, he espies the forces of the enemy on the other side of a considerable valley and rivulet. It was an affair of great danger to fight with such large forces in a disadvantageous situation. For the present, therefore, inasmuch as he knew that Cicero was released from the blockade, and thought that he might, on that account, relax his speed, he halted there and fortifies a camp in the most favourable position he can. And this, though it was small in itself, [there being] scarcely 7,000 men, and these too without baggage, still by the narrowness of the passages, he contracts as much as he can, with this object, that he may come into the greatest contempt with the enemy. In the meanwhile, scouts having been sent in all directions, he examines by what most convenient path he might cross the valley.

L. — That day, slight skirmishes of cavalry having taken place near the river, both armies kept in their own positions: the Gauls, because they were awaiting larger forces which had not then arrived; Caesar, [to see] if perchance by pretence of fear he could allure the enemy towards his position, so that he might engage in battle, in front of his camp, on this side of the valley; if he could not accomplish this, that, having inquired about the passes, he might cross the valley and the river with the less hazard. At day-break the cavalry of the enemy approaches to the camp and joins battle with our horse. Caesar orders the horse to give way purposely, and retreat to the camp: at the same time he orders the camp to be fortified with a higher rampart in all directions, the gates to be barricaded, and in executing these things as much confusion to be shown as possible, and to perform them under the pretence of fear.

LI. — Induced by all these things the enemy lead over their forces and draw up their line in a disadvantageous position; and as our men also had been led down from the ramparts, they approach nearer, and throw their weapons into the fortification from all sides, and sending heralds round, order it to be proclaimed that, if "any, either Gaul or Roman, was willing to go over to them before the third hour, it was permitted; after that time there would not be permission"; and so much did they disregard our men, that the gates having been blocked up with single rows of turf as a mere appearance, because they did not seem able to burst in that way, some began to pull down the rampart with their hands, others to fill up the trenches. Then Caesar, making a sally from all the gates, and sending out the cavalry, soon puts the enemy to flight, so that no one at all stood his ground with the intention of fighting; and he slew a great number of them, and deprived all of their arms.

LII. — Caesar, fearing to pursue them very far, because woods and morasses intervened, and also [because] he saw that they suffered no small loss in abandoning their position, reaches Cicero the same day with all his forces safe. He witnesses with surprise the towers, mantlets, and [other] fortifications belonging to the enemy: the legion having been drawn out, he finds that even every tenth soldier had not escaped without wounds. From all these things he judges with what danger and with what great courage matters had been conducted; he commends Cicero according to his desert and likewise the legion; he addresses individually the centurions and the tribunes of the soldiers, whose valour he had discovered to have been signal. He receives information of the death of Sabinus and Cotta from the prisoners. An assembly being held the following day, he states the occurrence; he consoles and encourages the soldiers; he suggests that the disaster, which had been occasioned by the misconduct and rashness of his lieutenant, should be borne with a patient mind, because by the favour of the immortal gods and their own valour, neither was lasting joy left to the enemy, nor very lasting grief to them.

LIII. — In the meanwhile the report respecting the victory of Caesar is conveyed to Labienus through the country of the Remi with incredible speed, so that, though he was about sixty miles distant from the winter-quarter of Cicero, and Caesar had arrived there after the ninth hour, before midnight a shout arose at the gates of the camp, by which shout an indication of the victory and a congratulation on the part of the Remi were given to Labienus. This report having been carried to the Treviri, Indutiormarus, who had resolved to attack the camp of Labienus the following day, flies by night and leads back all his forces into the country of the Treviri. Caesar sends back Fabius with his legion to his winter-quarters; he himself determines to winter with three legions near Samarobriva in three different quarters, and, because such great commotions had arisen in Gaul, he resolved to remain during the whole winter with the army himself. For the disaster respecting the death of Sabinus having been circulated among them, almost all the states of Gaul were deliberating about war, sending messengers and embassies into all quarters, inquiring what further measure they should take, and holding councils by night in secluded places. Nor did any period of the whole winter pass over without fresh anxiety to Caesar, or without his receiving some intelligence respecting the meetings and commotions of the Gauls. Among these, he is informed by L. Roscius, the lieutenant whom he had placed over the thirteenth legion, that large forces of those states of the

Gauls, which are called the Armoricae, had assembled for the purpose of attacking him and were not more than eight miles distant; but intelligence respecting the victory of Caesar being carried [to them], had retreated in such a manner that their departure appeared like a flight.

LIV. — But Caesar, having summoned to him the principal persons of each state, in one case by alarming them, since he declared that he knew what was going on, and in another case by encouraging them, retained a great part of Gaul in its allegiance. The Senones, however, which is a state eminently powerful and one of great influence among the Gauls, attempting by general design to slay Cavarinus whom Caesar had created king among them (whose brother, Moritasgus, had held the sovereignty at the period of the arrival of Caesar in Gaul, and whose ancestors had also previously held it) when he discovered their plot and fled, pursued him even to the frontiers [of the state], and drove him from his kingdom and his home; and, after having sent ambassadors to Caesar for the purpose of concluding a peace, when he ordered all their senate to come to him, did not obey that command. So far did it operate among those barbarian people, that there were found some to be the first to wage war; and so great a change of inclinations did it produce in all, that except the Aedui and the Remi, whom Caesar had always held in especial honour, the one people for their long standing and uniform fidelity towards the Roman people, the other for their late service in the Gallic war, there was scarcely a state which was not suspected by us. And I do not know whether that ought much to be wondered at, as well for several other reasons, as particularly because they who ranked above all nations for prowess in war, most keenly regretted that they had lost so much of that reputation as to submit to commands from the Roman people.

LV. — But the Treviri and Indutiomarus let no part of the entire winter pass without sending ambassadors across the Rhine, importuning the states, promising money, and asserting that, as a large portion of our army had been cut off, a much smaller portion remained. However, none of the German states could be induced to cross the Rhine, since "they had twice essayed it," they said, "in the war with Ariovistus and in the passage of the Tenchtheri there; that fortune was not to be tempted any more." Indutiomarus disappointed in this expectation, nevertheless began to raise troops, and discipline them, and procure horses from the neighbouring people and allure to him by great rewards the outlaws and convicts throughout Gaul. And such great influence had he already acquired for himself in Gaul by these means, that embassies were flocking to him in all directions, and seeking, publicly and privately, his favour and friendship.

LVI. — When he perceived that they were coming to him voluntarily; that on the one side the Senones and the Carnutes were stimulated by their consciousness of guilt, on the other side the Nervii and the Aduatuci were preparing war against the Romans, and that forces of volunteers would not be wanting to him if he began to advance from his own territories, he proclaims an armed council (this according to the custom of the Gauls is the commencement of war) at which, by a common law, all the youth were wont to assemble in arms; whoever of them comes last is killed in the sight of the whole assembly after being racked with every torture. In that council he declares Cingetorix, the leader of the other faction, his own son-in-law (whom we have above mentioned, as having embraced the protection of Caesar, and never having deserted him) an enemy and confiscates his property. When these things were finished, he asserts in the council that he, invited by the Senones and the Carnutes, and several other states of Gaul, was about to march thither through the territories of the Remi, devastate their lands, and attack the camp of Labienus: before he does that, he informs them of what he desires to be done.

LVII. — Labienus, since he was confining himself within a camp strongly fortified by the nature of the ground and by art, had no apprehensions as to his own and the legion's danger, but was devising that he might throw away no opportunity of conducting the war successfully. Accordingly, the speech of Indutiomarus, which he had delivered in the council, having been made known [to him] by Cingetorix and his allies, he sends messengers to the neighbouring states and summons horse from all quarters: he appoints to them a fixed day for assembling. In the meantime, Indutiomarus, with all his cavalry, nearly every day used to parade close to his [Labienus's] camp; at one time, that he might inform himself of the situation of the camp; at another time, for the purpose of conferring with or of intimidating him. Labienus confined his men within the fortifications and promoted the enemy's belief of his fear by whatever methods he could.

LVIII. — Since Indutiomarus was daily advancing up to the camp with greater defiance, all the cavalry of the neighbouring states which he [Labienus] had taken care to have sent for, having been admitted in one night, he confined all his men within the camp by guards with such great strictness, that that fact could by no means be reported or carried to the Treviri. In the meanwhile Indutiomarus, according to his daily practice, advances up to the camp and spends a great part of the day there: his horse cast their weapons, and with very insulting language call out our men to battle. No reply being given by our men, the enemy when they thought proper, depart towards evening in a disorderly and scattered manner, Labienus unexpectedly sends out all the cavalry by two gates; he gives this command and prohibition, that, when the enemy should be terrified and put to flight (which he foresaw would happen, as it did), they should all make for Indutiomarus, and no one wound any man before he should have seen him slain, because he was unwilling that he should escape, in consequence of gaining time by the delay [occasioned by the pursuit] of the rest. He offers great rewards for those who should kill him: he sends up the cohorts as a relief to the horse. The issue justifies the policy of the man, and, since all aimed at one, Indutiomarus is slain, having been overtaken at the very ford of the river, and his head is carried to the camp: the horse, when returning, pursue and slay all whom they can. This affair having been known, all the forces of the Eburones and the Nervii which had assembled, depart; and for a short time after this action, Caesar was less harassed in the government of Gaul.

Book VI

I. — Caesar, expecting for many reasons a greater commotion in Gaul, resolves to hold a levy by the means of M. Silanus, C. Antistius Reginus, and T. Sextius, his lieutenants: at the same time he requested of Cn. Pompey, the proconsul, that since he was remaining near the city invested with military command for the interests of the commonwealth, he would command those men whom when consul he had levied by the military oath in Cisalpine Gaul, to join their respective corps, and to proceed to him; thinking it of great importance, as far as regarded the opinion which the Gauls would entertain for the future, that the resources of Italy should appear so great, that if any loss should be sustained in war, not only could it be repaired in a short time, but likewise be further supplied by still larger forces. And when Pompey had granted this to the interests of the commonwealth and the claims of friendship, Caesar having quickly completed the levy by means of his lieutenants, after three legions had been both formed and brought to him before the winter [had] expired, and the number of those cohorts which he had lost under Q. Titurius had been doubled, taught the Gauls, both by his dispatch and by his forces, what the discipline and the power of the Roman people could accomplish.

II. — Indutiomarus having been slain, as we have stated, the government was conferred upon his relatives by the Treviri. They cease not to importune the neighbouring Germans and to promise them money: when they could not obtain [their object] from those nearest them, they try those more remote. Having found some states willing to accede to their wishes, they enter into a compact with them by a mutual oath, and give hostages as a security for the money: they attach Ambiorix to them by an alliance and confederacy. Caesar, on being informed of their acts, since he saw that war was being prepared on all sides, that the Nervii, Aduatuci, and Menapii, with the addition of all the Germans on this side of the Rhine were under arms, that the Senones did not assemble according to his command, and were concerting measures with the Carnutes and the neighbouring states, that the Germans were importuned by the Treviri in frequent embassies, thought that he ought to take measures for the war earlier [than usual].

III. — Accordingly, while the winter was not yet ended, having concentrated the four nearest legions, he marched unexpectedly into the territories of the Nervii, and before they could either assemble, or retreat, after capturing a large number of cattle and of men, and wasting their lands and giving up that booty to the soldiers, compelled them to enter into a surrender and give him hostages. That business having been speedily executed, he again led his legions back into winter-quarters. Having proclaimed a council of Gaul in the beginning of the spring, as he had been accustomed [to do], when the deputies from the rest, except the Senones, the Carnutes, and the Treviri, had come, judging this to be the commencement of war and revolt, that he might appear to consider all things of less consequence [than that war], he transfers the council to Lutetia of the Parisii. These were adjacent to the Senones, and had united their state to them during the memory of their fathers, but were thought to have no part in the present plot. Having proclaimed this from the tribunal, he advances the same day towards the Senones with his legions and arrives among them by long marches.

IV. — Acco, who had been the author of that enterprise, on being informed of his arrival, orders the people to assemble in the towns; to them, while attempting this and before it could be accomplished, news is brought that the Romans are close at hand: through necessity they give over their design and send ambassadors to Caesar for the purpose of imploring pardon; they make advances to him through the Aedui, whose state was from ancient times under the protection of Rome. Caesar readily grants them pardon and receives their excuse at the request of the Aedui; because he thought that the summer season was one for an impending war, not for an investigation. Having imposed one hundred hostages, he delivers these to the Aedui to be held in charge by them. To the same place the Carnutes send ambassadors and hostages, employing as their mediators the Remi, under whose protection they were: they receive the same answers. Caesar concludes the council and imposes a levy of cavalry on the states.

V. — This part of Gaul having been tranquillized, he applies himself entirely both in mind and soul to the war with the Treviri and Ambiorix. He orders Cavarinus to march with him with the cavalry of the Senones, lest any commotion should arise either out of his hot temper, or out of the hatred of the state which he had incurred. After arranging these things, as he considered it certain that Ambiorix would not contend in battle, he watched his other plans attentively. The Menapii bordered on the territories of the Eburones, and were protected by one continued extent of morasses and woods; and they alone out of Gaul had never sent ambassadors to Caesar on the subject of peace. Caesar knew that a tie of hospitality subsisted between them and Ambiorix: he also discovered that the latter had entered into an alliance with the Germans by means of the Treviri. He thought that these auxiliaries ought to be detached from him before he provoked him to war; lest he, despairing of safety, should either proceed to conceal himself in the territories of the Menapii, or should be driven to coalesce with the Germans beyond the Rhine. Having entered upon this resolution, he sends the baggage of the whole army to Labienus, in the territories of the Treviri and orders two legions to proceed to him: he himself proceeds against the Menapii with five lightly-equipped legions. They, having assembled no troops, as they relied on the defence of their position, retreat into the woods and morasses, and convey thither all their property.

VI. — Caesar, having divided his forces with C. Fabius, his lieutenant, and M. Crassus, his questor, and having hastily constructed some bridges, enters their country in three divisions, burns their houses and villages, and gets possession of a large number of cattle and men. Constrained by these circumstances, the Menapii send ambassadors to him for the purpose of suing for peace. He, after receiving hostages, assures them that he will consider them in the number of his enemies if they shall receive within their territories either Ambiorix or his ambassadors. Having determinately settled these things, he left among the Menapii, Commius the Atrebatian with some cavalry as a guard; he himself proceeds toward the Treviri.

VII. — While these things are being performed by Caesar, the Treviri, having drawn together large forces of infantry and of cavalry, were preparing to attack Labienus and the legion which was wintering in their territories, and were already not further distant from him than a journey of two days, when they learn that two legions had arrived by the order of Caesar. Having pitched their camp fifteen miles off, they resolve to await the support of the Germans. Labienus, having learned the design of the enemy, hoping that through their rashness there would be some opportunity of engaging, after leaving a guard of five cohorts for the baggage, advances against the enemy with twenty-five cohorts and a large body of cavalry, and, leaving the space of a mile between them, fortifies his camp. There was between Labienus and the enemy a river difficult to cross and with steep banks: this neither did he himself design to cross, nor did he suppose the enemy would cross it. Their hope of auxiliaries was daily increasing. He [Labienus] openly says in a council that "since the Germans are said to be approaching, he would not bring into uncertainty his own and the army's fortunes, and the next day would move his camp at early dawn. These words are quickly carried to the enemy, since out of so large a number of cavalry composed of Gauls, nature compelled some to favour the Gallic interests. Labienus, having assembled the tribunes of the soldiers and principal centurions by night, states what his design is, and, that he may the more easily give the enemy a belief of his fears, he orders the camp to be moved with greater noise and confusion than was usual with the Roman people. By these means he makes his departure [appear], like a retreat. These things, also, since the camps were so near, are reported to the enemy by scouts before daylight.

VIII. — Scarcely had the rear advanced beyond the fortifications when the Gauls, encouraging one another "not to cast from their hands the anticipated booty, that it was a tedious thing, while the Romans were panic stricken, to be waiting for the aid of the Germans, and that their dignity did not suffer them to fear to attack with such great forces so small a band, particularly when retreating and encumbered," do not hesitate to cross the river and give battle in a disadvantageous position. Labienus suspecting that these things would happen, was proceeding quietly, and using the same pretence of a march, in order that he might entice them across the river. Then, having sent forward the baggage some short distance and placed it on a certain eminence, he says, "Soldiers, you have the opportunity you have sought: you hold the enemy in an encumbered and disadvantageous position: display to us your leaders the same valour you have ofttimes displayed to your general: imagine that he is present and actually sees these exploits." At the same time he orders the troops to face about towards the enemy and form in line of battle, and, despatching a few troops of cavalry as a guard for the baggage, he places the rest of the horse on the wings. Our men, raising a shout, quickly throw their javelins at the enemy. They, when, contrary to their expectation, they saw those whom they believed to be retreating, advance towards them with threatening banners, were not able to sustain even the charge, and, being put to flight at the first onslaught, sought the nearest woods: Labienus pursuing them with the cavalry, upon a large number being slain, and several taken prisoners, got possession of the state a few days after; for the Germans who were coming to the aid of the Treviri, having been informed of their flight, retreated to their homes. The relations of Indutiomarus, who had been the promoters of the revolt, accompanying them, quitted their own state with them. The supreme power and government were delivered to Cingetorix, whom we have stated to have remained firm in his allegiance from the commencement.

IX. — Caesar, after he came from the territories of the Menapii into those of the Treviri, resolved for two reasons to cross the Rhine; one of which was, because they had sent assistance to the Treviri against him; the other, that Ambiorix might not have a retreat among them. Having determined on these matters, he began to build a bridge a little above that place, at which he had before conveyed over his army. The plan having been known and laid down, the work is accomplished in a few days by the great exertion of the soldiers. Having left a strong guard at the bridge on the side of the Treviri, lest any commotion should suddenly arise among them, he leads over the rest of the forces and the cavalry. The Ubii, who before had sent hostages and come to a capitulation, send ambassadors to him, for the purpose of vindicating themselves, to assure him that "neither had auxiliaries been sent to the Treviri from their state, nor had they violated their allegiance"; they entreat and beseech him "to spare them, lest, in his common hatred of the Germans, the innocent should suffer the penalty of the guilty: they promise to give more hostages, if he desire them." Having investigated the case, Caesar finds that the auxiliaries had been sent by the Suevi; he accepts the apology of the Ubii, and makes minute inquiries concerning the approaches and the routes to the territories of the Suevi. X. — In the meanwhile he is informed by the Ubii, a few days after, that the Suevi are drawing all their forces into one place, and are giving orders to those nations which are under their government to send auxiliaries of infantry and of cavalry. Having learned these things, he provides a supply of corn, selects a proper place for his camp, and commands the Ubii to drive off their cattle and carry away all their possessions from the country parts into the towns, hoping that they, being a barbarous and ignorant people, when harassed by the want of provisions, might be brought to an engagement on disadvantageous terms: he orders them to send numerous scouts among the Suevi, and learn what things are going on among them. They execute the orders, and, a few days having intervened, report that all the Suevi, after certain intelligence concerning the army of the Romans had come, retreated with all their own forces and those of their allies, which they had assembled, to the utmost extremities of their territories: that there is a wood there of very great extent, which is called Bacenis; that this stretches a great way into the interior, and, being opposed as a natural barrier, defends from injuries and incursions the Cherusci against the Suevi, and the Suevi against the Cherusci: that at the entrance of that forest the Suevi had determined to await the coming up of the Romans.

XI. — Since we have come to this place, it does not appear to be foreign to our subject to lay before the reader an account of the manners of Gaul and Germany, and wherein these nations differ from each other. In Gaul there are factions not only in all the states, and in all the cantons and their divisions, but almost in each family, and of these factions those are the leaders who are considered according to their judgment to possess the greatest influence, upon whose will and determination the management of all affairs and measures depends. And that seems to have been instituted in ancient times with this view, that no one of the common people should be in want of support against one more powerful; for none [of those leaders] suffers his party to be oppressed and defrauded, and if he do otherwise, he has no influence among his party. This same policy exists throughout the whole of Gaul; for all the states are divided into two factions.

XII. — When Caesar arrived in Gaul, the Aedui were the leaders of one faction, the Sequani of the other. Since the latter were less powerful by themselves, inasmuch as the chief influence was from of old among the Aedui, and their dependencies were great, they had united to themselves the Germans and Ariovistus, and had brought them over to their party by great sacrifices and promises. And having fought several successful battles and slain all the nobility of the Aedui, they had so far surpassed them in power, that they brought over, from the Aedui to themselves, a large portion of their dependants and received from them the sons of their leading men as hostages, and compelled them to swear in their public character that they would enter into no design against them; and held a portion of the neighbouring land, seized on by force, and possessed the sovereignty of the whole of Gaul. Divitiacus urged by this necessity, had proceeded to Rome to the senate, for the purpose of entreating assistance, and had returned without accomplishing his object. A change of affairs ensued on the arrival of Caesar, the hostages were returned to the Aedui, their old dependencies restored, and new acquired through Caesar (because those who had attached themselves to their alliance saw that they enjoyed a better state and a milder government), their other interests, their influence, their reputation were likewise increased, and in consequence, the Sequani lost the sovereignty. The Remi succeeded to their place, and, as it was perceived that they equalled the Aedui in favour with Caesar, those, who on account of their old animosities could by no means coalesce with the Aedui, consigned themselves in clientship to the Remi. The latter carefully protected them. Thus they possessed both a new and suddenly acquired influence. Affairs were then in that position, that the Aedui were considered by far the leading people, and the Remi held the second post of honour.

XIII. — Throughout all Gaul there are two orders of those men who are of any rank and dignity: for the commonality is held almost in the condition of slaves, and dares to undertake nothing of itself and is admitted to no deliberation. The greater part, when they are pressed either by debt, or the large amount of their tributes, or the oppression of the more powerful, give themselves up in vassalage to the nobles, who possess over them the same rights without exception as masters over their slaves. But of these two orders, one is that of the Druids, the other that of the knights. The former are engaged in things sacred, conduct the public and the private sacrifices, and interpret all matters of religion. To these a large number of the young men resort for the purpose of instruction, and they [the Druids] are in great honour among them. For they determine respecting almost all controversies, public and private; and if any crime has been perpetrated, if murder has been committed, if there be any dispute about an inheritance, if any about boundaries, these same persons decide it; they decree rewards and punishments if any one, either in a private or public capacity, has not submitted to their decision, they interdict him from the sacrifices. This among them is the most heavy punishment. Those who have been thus interdicted are esteemed in the number of the impious and the criminal: all shun them, and avoid their society and conversation, lest they receive some evil from their contact; nor is justice administered to them when seeking it, nor is any dignity bestowed on them. Over all these Druids one presides, who possesses supreme authority among them. Upon his death, if any individual among the rest is pre-eminent in dignity, he succeeds; but, if there are many equal, the election is made by the suffrages of the Druids; sometimes they even contend for the presidency with arms. These assemble at a fixed period of the year in a consecrated place in the territories of the Carnutes, which is reckoned the central region of the whole of Gaul. Hither all, who have disputes, assemble from every part, and submit to their decrees and determinations. This institution is supposed to have been devised in Britain, and to have been brought over from it into Gaul; and now those who desire to gain a more accurate knowledge of that system generally proceed thither for the purpose of studying it.

XIV. — The Druids do not go to war, nor pay tribute together with the rest; they have an exemption from military service and a dispensation in all matters. Induced by such great advantages, many embrace this profession of their own accord, and [many] are sent to it by their parents and relations. They are said there to learn by heart a great number of verses; accordingly some remain in the course of training twenty years. Nor do they regard it lawful to commit these to writing, though in almost all other matters, in their public and private transactions, they use Greek characters. That practice they seem to me to have adopted for two reasons; because they neither desire their doctrines to be divulged among the mass of the people, nor those who learn, to devote themselves the less to the efforts of memory, relying on writing; since it generally occurs to most men, that, in their dependence on writing, they relax their diligence in learning thoroughly, and their employment of the memory. They wish to inculcate this as one of their leading tenets, that souls do not become extinct, but pass after death from one body to another, and they think that men by this tenet are in a great degree excited to valour, the fear of death being disregarded. They likewise discuss and impart to the youth many things respecting the stars and their motion, respecting the extent of the world and of our earth, respecting the nature of things, respecting the power and the majesty of the immortal gods.

XV. — The other order is that of the knights. These, when there is occasion and any war occurs (which before Caesar's arrival was for the most part wont to happen every year, as either they on their part were inflicting injuries or repelling those which others inflicted on them), are all engaged in war. And those of them most distinguished by birth and resources, have the greatest number of vassals and dependants about them. They acknowledge this sort of influence and power only.

XVI. — The nation of all the Gauls is extremely devoted to superstitious rites; and on that account they who are troubled with unusually severe diseases and they who are engaged in battles and dangers, either sacrifice men as victims, or vow that they will sacrifice them, and employ the Druids as the performers of those sacrifices; because they think that unless the life of a man be offered for the life of a man, the mind of the immortal gods cannot be rendered propitious, and they have sacrifices of that kind ordained for national purposes. Others have figures of vast size, the limbs of which formed of osiers they fill with living men, which being set on fire, the men perish enveloped in the flames. They consider that the oblation of such as have been taken in theft, or in robbery, or any other offence, is more acceptable to the immortal gods; but when a supply of that class is wanting, they have recourse to the oblation of even the innocent.

XVII. — They worship as their divinity, Mercury in particular, and have many images of him, and regard him as the inventor of all arts, they consider him, the guide of their journeys and marches, and believe him to have very great influence over the acquisition of gain and mercantile transactions. Next to him they worship Apollo, and Mars, and Jupiter, and Minerva; respecting these deities they have for the most part the same belief as other nations: that Apollo averts diseases, that Minerva imparts the invention of manufactures, that Jupiter possesses the sovereignty of the heavenly powers; that Mars presides over wars. To him when they have determined to engage in battle, they commonly vow those things they shall take in war. When they have conquered, they sacrifice whatever captured animals may have survived the conflict, and collect the other things into one place. In many states you may see piles of these things heaped up in their consecrated spots; nor does it often happen that any one, disregarding the sanctity of the case, dares either to secrete in his house things captured, or take away those deposited; and the most severe punishment, with torture, has been established for such a deed.

XVIII. — All the Gauls assert that they are descended from the god Dis, and say that this tradition has been handed down by the Druids. For that reason they compute the divisions of every season, not by the number of days, but of nights; they keep birthdays and the beginnings of months and years in such an order that the day follows the night. Among the other usages of their life, they differ in this from almost all other nations, that they do not permit their children to approach them openly until they are grown up so as to be able to bear the service of war; and they regard it as indecorous for a son of boyish age to stand in public in the presence of his father.

XIX. — Whatever sums of money the husbands have received in the name of dowry from their wives, making an estimate of it, they add the same amount out of their own estates. An account is kept of all this money conjointly, and the profits are laid by: whichever of them shall have survived [the other], to that one the portion of both reverts, together with the profits of the previous time. Husbands have power of life and death over their wives as well as over their children: and when the father of a family, born in a more than commonly distinguished rank, has died, his relations assemble, and, if the circumstances of his death are suspicious, hold an investigation upon the wives in the manner adopted towards slaves; and if proof be obtained, put them to severe torture, and kill them. Their funerals, considering the state of civilization among the Gauls, are magnificent and costly; and they cast into the fire all things, including living creatures, which they suppose to have been dear to them when alive; and, a little before this period, slaves and dependants, who were ascertained to have been beloved by them, were, after the regular funeral rites were completed, burnt together with them.

XX. — Those states which are considered to conduct their commonwealth more judiciously, have it ordained by their laws, that, if any person shall have heard by rumour and report from his neighbours anything concerning the commonwealth, he shall convey it to the magistrate and not impart it to any other; because it has been discovered that inconsiderate and inexperienced men were often alarmed by false reports and driven to some rash act, or else took hasty measures in affairs of the highest importance. The magistrates conceal those things which require to be kept unknown; and they disclose to the people whatever they determine to be expedient. It is not lawful to speak of the commonwealth, except in council.

XXI. — The Germans differ much from these usages, for they have neither Druids to preside over sacred offices, nor do they pay great regard to sacrifices. They rank in the number of the gods those alone whom they behold, and by whose instrumentality they are obviously benefited, namely, the sun, fire, and the moon; they have not heard of the other deities even by report. Their whole life is occupied in hunting and in the pursuits of the military art; from childhood they devote themselves to fatigue and hardships. Those who have remained chaste for the longest time, receive the greatest commendation among their people: they think that by this the growth is promoted, by this the physical powers are increased and the sinews are strengthened. And to have had knowledge of a woman before the twentieth year they reckon among the most disgraceful acts; of which matter there is no concealment, because they bathe promiscuously in the rivers and [only] use skins or small cloaks of deers' hides, a large portion of the body being in consequence naked.

XXII. — They do not pay much attention to agriculture, and a large portion of their food consists in milk, cheese, and flesh; nor has any one a fixed quantity of land or his own individual limits; but the magistrates and the leading men each year apportion to the tribes and families, who have united together, as much land as, and in the place in which, they think proper, and the year after compel them to remove elsewhere. For this enactment they advance many reasons — lest seduced by long-continued custom, they may exchange their ardour in the waging of war for agriculture; lest they may be anxious to acquire extensive estates, and the more powerful drive the weaker from their possessions; lest they construct their houses with too great a desire to avoid cold and heat; lest the desire of wealth spring up, from which cause divisions and discords arise; and that they may keep the common people in a contented state of mind, when each sees his own means placed on an equality with [those of] the most powerful.

XXIII. — It is the greatest glory to the several states to have as wide deserts as possible around them, their frontiers having been laid waste. They consider this the real evidence of their prowess, that their neighbours shall be driven out of their lands and abandon them, and that no one dare settle near them; at the same time they think that they shall be on that account the more secure, because they have removed the apprehension of a sudden incursion. When a state either repels war waged against it, or wages it against another, magistrates are chosen to preside over that war with such authority, that they have power of life and death. In peace there is no common magistrate, but the chiefs of provinces and cantons administer justice and determine controversies among their own people. Robberies which are committed beyond the boundaries of each state bear no infamy, and they avow that these are committed for the purpose of disciplining their youth and of preventing sloth. And when any of their chiefs has said in an assembly "that he will be their leader, let those who are willing to follow, give in their names"; they who approve of both the enterprise and the man arise and promise their assistance and are applauded by the people; such of them as have not followed him are accounted in the number of deserters and traitors, and confidence in all matters is afterwards refused them. To injure guests they regard as impious; they defend from wrong those who have come to them for any purpose whatever, and esteem them inviolable; to them the houses of all are open and maintenance is freely supplied.

XXIV. — And there was formerly a time when the Gauls excelled the Germans in prowess, and waged war on them offensively, and, on account of the great number of their people and the insufficiency of their land, sent colonies over the Rhine. Accordingly, the Volcae Tectosages seized on those parts of Germany which are the most fruitful [and lie] around the Hercynian forest (which, I perceive, was known by report to Eratosthenes and some other Greeks, and which they call Orcynia) and settled there. Which nation to this time retains its position in those settlements, and has a very high character for justice and military merit: now also they continue in the same scarcity, indigence, hardihood, as the Germans, and use the same food and dress; but their proximity to the Province and knowledge of commodities from countries beyond the sea supplies to the Gauls many things tending to luxury as well as civilization. Accustomed by degrees to be overmatched and worsted in many engagements, they do not even compare themselves to the Germans in prowess.

XXV.—The breadth of this Hercynian forest, which has been referred to above, is to a quick traveller, a journey of nine days. For it cannot be otherwise computed, nor are they acquainted with the measures of roads. It begins at the frontiers of the Helvetii, Nemetes, and Rauraci, and extends in a right line along the river Danube to the territories of the Daci and the Anartes: it bends thence to the left in a different direction from the river, and owing to its extent touches the confines of many nations; nor is there any person belonging to this part of Germany who says that he either has gone to the extremity of that forest, though he had advanced a journey of sixty days, or has heard in what place it begins. It is certain that many kinds of wild beasts are produced in it which have not been seen in other parts; of which the following are such as differ principally from other animals, and appear worthy of being committed to record.

XXVI. — There is an ox of the shape of a stag, between whose ears a horn rises from the middle of the forehead, higher and straighter than those horns which are known to us. From the top of this, branches, like palms; stretch out a considerable distance. The shape of the female and of the male is the same; the appearance and the size of the horns is the same.

XXVII. — There are also [animals] which are called elks. The shape of these, and the varied colour of their skins, is much like roes, but in size they surpass them a little and are destitute of horns, and have legs without joints and ligatures; nor do they lie down for the purpose of rest, nor, if they have been thrown down by any accident, can they raise or lift themselves up. Trees serve as beds to them; they lean themselves against them, and thus reclining only slightly, they take their rest; when the huntsmen have discovered from the footsteps of these animals whither they are accustomed to betake themselves, they either undermine all the trees at the roots, or cut into them so far that the upper part of the trees may appear to be left standing. When they have leant upon them, according to their habit, they knock down by their weight the unsupported trees, and fall down themselves along with them.

XXVIII. — There is a third kind, consisting of those animals which are called uri. These are a little below the elephant in size, and of the appearance, colour, and shape of a bull. Their strength and speed are extraordinary; they spare neither man nor wild beast which they have espied. These the Germans take with much pains in pits and kill them. The young men harden themselves with this exercise, and practice themselves in this kind of hunting, and those who have slain the greatest number of them, having produced the horns in public, to serve as evidence, receive great praise. But not even when taken very young can they be rendered familiar to men and tamed. The size, shape, and appearance of their horns differ much from the horns of our oxen. These they anxiously seek after, and bind at the tips with silver, and use as cups at their most sumptuous entertainments.

XXIX. — Caesar, after he discovered through the Ubian scouts that the Suevi had retired into their woods, apprehending a scarcity of corn, because, as we have observed above, all the Germans pay very little attention to agriculture, resolved not to proceed any farther; but, that he might not altogether relieve the barbarians from the fear of his return, and that he might delay their succours, having led back his army, he breaks down, to the length of 200 feet, the farther end of the bridge, which joined the banks of the Ubii, and, at the extremity of the bridge raises towers of four stories, and stations a guard of twelve cohorts for the purpose of defending the bridge, and strengthens the place with considerable fortifications. Over that fort and guard he appointed C. Volcatius Tullus, a young man; he himself, when the corn began to ripen, having set forth for the war with 40 Ambiorix (through the forest Arduenna, which is the largest of all Gaul, and reaches from the banks of the Rhine and the frontiers of the Treviri to those of the Nervii, and extends over more than 500 miles), he sends forward L. Minucius Basilus with all the cavalry, to try if he might gain any advantage by rapid marches and the advantage of time, he warns him to forbid fires being made in the camp, lest any indication of his approach be given at a distance: he tells him that he will follow immediately.

XXX. — Basilus does as he was commanded; having performed his march rapidly, and even surpassed the expectations of all, he surprises in the fields many not expecting him; through their information he advances towards Ambiorix himself, to the place in which he was said to be with a few horse. Fortune accomplishes much, not only in other matters, but also in the art of war. For as it happened by a remarkable chance, that he fell upon [Ambiorix] himself unguarded and unprepared, and that his arrival was seen by the people before the report or information of his arrival was carried thither; so it was an incident of extraordinary fortune that, although every implement of war which he was accustomed to have about him was seized, and his chariots and horses surprised, yet he himself escaped death. But it was effected owing to this circumstance, that his house being surrounded by a wood, (as are generally the dwellings of the Gauls, who, for the purpose of avoiding heat, mostly seek the neighbourhood of woods and rivers) his attendants and friends in a narrow spot sustained for a short time the attack of our horse. While they were fighting, one of his followers mounted him on a horse: the woods sheltered him as he fled. Thus fortune tended much both towards his encountering and his escaping danger.

XXXI. — Whether Ambiorix did not collect his forces from cool deliberation, because he considered he ought not to engage in a battle, or [whether] he was debarred by time and prevented by the sudden arrival of our horse, when he supposed the rest of the army was closely following, is doubtful; but certainly, despatching messengers through the country, he ordered every one to provide for himself; and a part of them fled into the forest Arduenna, a part into the extensive morasses; those who were nearest the ocean, concealed themselves in the islands which the tides usually form; many, departing from their territories, committed themselves and all their possessions to perfect strangers. Cativolcus, king of one-half of the Eburones, who had entered into the design together with Ambiorix, since, being now worn out by age, he was unable to endure the fatigue either of war or flight, having cursed Ambiorix with every imprecation, as the person who had been the contriver of that measure, destroyed himself with the juice of the yew tree, of which there is a great abundance in Gaul and Germany.

XXXII. — The Segui and Condrusi, of the nation and number of the Germans, and who are between the Eburones and the Treviri, sent ambassadors to Caesar to entreat that he would not regard them in the number of his enemies, nor consider that the cause of all the Germans on this side the Rhine was one and the same; that they had formed no plans of war, and had sent no auxiliaries to Ambiorix. Caesar, having ascertained this fact by an examination of his prisoners commanded that if any of the Eburones in their flight had repaired to them, they should be sent back to him; he assures them that if they did that, he will not injure their territories. Then, having divided his forces into three parts, he sent the baggage of all the legions to Aduatuca. That is the name of a fort. This is nearly in the middle of the Eburones, where Titurius and Aurunculeius had been quartered for the purpose of wintering. This place he selected as well on other accounts as because the fortifications of the previous year remained, in order that he might relieve the labour of the soldiers. He left the fourteenth legion as a guard for the baggage, one of those three which he had lately raised in Italy and brought over. Over that legion and camp he places Q. Tullius Cicero and gives him 200 horse.

XXXIII. — Having divided the army, he orders T. Labienus to proceed with three legions towards the ocean into those parts which border on the Menappii; he sends C. Trebonius with a like number of legions to lay waste that district which lies contiguous to the Aduatuci; he himself determines to go with the remaining three to the river Sambre, which flows into the Meuse, and to the most remote parts of Arduenna, whither he heard that Ambiorix had gone with a few horse. When departing, he promises that he will return before the end of the seventh day, on which day he was aware corn was due to that legion which was being left in garrison. He directs Labienus and Trebonius to return by the same day, if they can do so agreeably to the interests of the republic; so that their measures having been mutually imparted, and the plans of the enemy having been discovered, they might be able to commence a different line of operations.

XXXIV. — There was, as we have above observed, no regular army, nor a town, nor a garrison which could defend itself by arms; but the people were scattered in all directions. Where either a hidden valley, or a woody spot, or a difficult morass furnished any hope of protection or of security to any one, there he had fixed himself. These places were known to those that dwelt in the neighbourhood, and the matter demanded great attention, not so much in protecting the main body of the army (for no peril could occur to them altogether from those alarmed and scattered troops), as in preserving individual soldiers; which in some measure tended to the safety of the army. For both the desire of booty was leading many too far, and the woods with their unknown and hidden routes would not allow them to go in large bodies. If he desired the business to be completed and the race of those infamous people to be cut off, more bodies of men must be sent in several directions and the soldiers must be detached on all sides; if he were disposed to keep the companies at their standards, as the established discipline and practice of the Roman army required, the situation itself was a safeguard to the barbarians, nor was there wanting to individuals the daring to lay secret ambuscades and beset scattered soldiers. But amidst difficulties of this nature as far as precautions could be taken by vigilance, such precautions were taken; so that some opportunities of injuring the enemy were neglected, though the minds of all were burning to take revenge, rather than that injury should be effected with any loss to our soldiers. Caesar despatches messengers to the neighbouring states; by the hope of booty he invites all to him, for the purpose of plundering the Eburones, in order that the life of the Gauls might be hazarded in the woods rather than the legionary soldiers; at the same time, in order that a large force being drawn around them, the race and name of that state may be annihilated for such a crime. A large number from all quarters speedily assembles.

XXXV. — These things were going on in all parts of the territories of the Eburones, and the seventh day was drawing near, by which day Caesar had purposed to return to the baggage and the legion. Here it might be learned how much fortune achieves in war, and how great casualties she produces. The enemy having been scattered and alarmed, as we related above, there was no force which might produce even a slight occasion of fear. The report extends beyond the Rhine to the Germans that the Eburones are being pillaged, and that all were without distinction invited to the plunder. The Sigambri, who are nearest to the Rhine, by whom, we have mentioned above, the Tenchtheri and Usipetes were received after their retreat, collect 2,000 horse; they cross the Rhine in ships and barks thirty miles below that place where the bridge was entire and the garrison left by Caesar; they arrive at the frontiers of the Eburones, surprise many who were scattered in flight, and get possession of a large amount of cattle, of which barbarians are extremely covetous. Allured by booty, they advance farther; neither morass nor forest obstructs these men, born amidst war and depredations; they inquire of their prisoners in what parts Caesar is; they find that he has advanced farther, and learn that all the army has removed. Thereon one of the prisoners says, "Why do you pursue such wretched and trifling spoil; you, to whom it is granted to become even now most richly endowed by fortune? In three hours you can reach Aduatuca; there the Roman army has deposited all its fortunes; there is so little of a garrison that not even the wall can be manned, nor dare any one go beyond the fortifications." A hope having been presented them, the Germans leave in concealment the plunder they had acquired; they themselves hasten to Aduatuca, employing as their guide the same man by whose information they had become informed of these things.

XXXVI. — Cicero, who during all the foregoing days had kept his soldiers in camp with the greatest exactness, and agreeably to the injunctions of Caesar, had not permitted even any of the camp-followers to go beyond the fortification, distrusting on the seventh day that Caesar would keep his promise as to the number of days, because he heard that he had proceeded farther, and no report as to his return was brought to him, and being urged at the same time by the expressions of those who called his tolerance almost a siege, if, forsooth, it was not permitted them to go out of the camp, since he might expect no disaster, whereby he could be injured, within three miles of the camp, while nine legions and all the cavalry were under arms, and the enemy scattered and almost annihilated, sent five cohorts into the neighbouring cornlands, between which and the camp only one hill intervened, for the purpose of foraging. Many soldiers of the legions had been left invalided in the camp, of whom those who had recovered in this space of time, being about 300, are set together under one standard; a large number of soldiers' attendants besides, with a great number of beasts of burden, which had remained in the camp, permission being granted, follow them.

XXXVII. — At this very time, the German horse by chance come up, and immediately, with the same speed with which they had advanced, attempt to force the camp at the Decuman gate, nor were they seen, in consequence of woods lying in the way on that side, before they were just reaching the camp: so much so, that the sutlers who had their booths under the rampart had not an opportunity of retreating within the camp. Our men, not anticipating it, are perplexed by the sudden affair, and the cohort on the outpost scarcely sustains the first attack. The enemy spread themselves on the other sides to ascertain if they could find any access. Our men with difficulty defend the gates; the very position of itself and the fortification secures the other accesses. There is a panic in the entire camp, and one inquires of another the cause of the confusion, nor do they readily determine whither the standards should be borne, nor into what quarter each should betake himself. One avows that the camp is already taken, another maintains that, the enemy having destroyed the army and commander-in-chief, are come thither as conquerors; most form strange superstitious fancies from the spot, and place before their eyes the catastrophe of Cotta and Titurius, who had fallen in the same fort. All being greatly disconcerted by this alarm, the belief of the barbarians is strengthened that there is no garrison within, as they had heard from their prisoner. They endeavour to force an entrance and encourage one another not to cast from their hands so valuable a prize.

XXXVIII.— P. Sextius Baculus, who had led a principal century under Caesar (of whom we have made mention in previous engagements), had been left an invalid in the garrison, and had now been five days without food. He, distrusting his own safety and that of all, goes forth from his tent unarmed; he sees that the enemy are close at hand and that the matter is in the utmost danger; he snatches arms from those nearest, and stations himself at the gate. The centurions of that cohort which was on guard follow him; for a short time they sustain the fight together. Sextius faints, after receiving many wounds; he is with difficulty saved, drawn away by the hands of the soldiers. This space having intervened, the others resume courage, so far as to venture to take their place on the fortifications and present the aspect of defenders.

XXXIX. — The foraging having in the meantime been completed, our soldiers distinctly hear the shout; the horse hasten on before and discover in what danger the affair is. But here there is no fortification to receive them, in their alarm: those last enlisted and unskilled in military discipline turn their faces to the military tribune and the centurions; they wait to find what orders may be given by them. No one is so courageous as not to be disconcerted by the suddenness of the affair. The barbarians, espying our standard in the distance, desist from the attack; at first they suppose that the legions, which they had learned from their prisoners had removed farther off, had returned; afterwards, despising their small number, they make an attack on them at all sides.

XL. — The camp-followers run forward to the nearest rising ground; being speedily driven from this they throw themselves among the standards and companies: they thus so much the more alarm the soldiers already affrighted. Some propose that, forming a wedge, they suddenly break through, since the camp was so near; and if any part should be surrounded and slain, they fully trust that at least the rest may be saved; others, that they take their stand on an eminence, and all undergo the same destiny. The veteran soldiers, whom we stated to have set out together [with the others] under a standard, do not approve of this. Therefore encouraging each other, under the conduct of Caius Trebonius, a Roman knight, who had been appointed over them, they break through the midst of the enemy, and arrive in the camp safe to a man. The camp-attendants and the horse following close upon them with the same impetuosity, are saved by the courage of the soldiers. But those who had taken their stand upon the eminence having even now acquired no experience of military matters, neither could persevere in that resolution which they approved of, namely, to defend themselves from their higher position, nor imitate that vigour and speed which they had observed to have availed others; but, attempting to reach the camp, had descended into an unfavourable situation. The centurions, some of whom had been promoted for their valour from the lower ranks of other legions to higher ranks in this legion, in order that they might not forfeit their glory for military exploits previously acquired, fell together fighting most valiantly. The enemy having been dislodged by their valour, a part of the soldiers arrived safe in camp contrary to their expectations; a part perished, surrounded by the barbarians.

XLI.—The Germans, despairing of taking the camp by storm, because they saw that our men had taken up their position on the fortifications, retreated beyond the Rhine with that plunder which they had deposited in the woods. And so great was the alarm, even after the departure of the enemy, that when C. Volusenus, who had been sent with the cavalry, arrived that night, he could not gain credence that Caesar was close at hand with his army safe. Fear had so pre-occupied the minds of all, that, their reason being almost estranged, they said that all the other forces having been cut off, the cavalry alone had arrived there by flight, and asserted that, if the army were safe, the Germans would not have attacked the camp: which fear the arrival of Caesar removed.

XLII. — He, on his return, being well aware of the casualties of war, complained of one thing [only], namely, that the cohorts had been sent away from the outposts and garrison [duty], and pointed out that room ought not to have been left for even the most trivial casualty; that fortune had exercised great influence in the sudden arrival of their enemy; much greater, in that she had turned the barbarians away from the very rampart and gates of the camp. Of all which events, it seemed the most surprising that the Germans, who had crossed the Rhine with this object, that they might plunder the territories of Ambiorix, being led to the camp of the Romans, rendered Ambiorix a most acceptable service.

XLIII. — Caesar, having again marched to harass the enemy, after collecting a large number [of auxiliaries] from the neighbouring states, despatches them in all directions. All the villages and all the buildings, which each beheld, were on fire: spoil was being driven off from all parts; the corn not only was being consumed by so great numbers of cattle and men, but also had fallen to the earth, owing to the time of the year and the storms; so that if any had concealed themselves for the present, still, it appeared likely that they must perish through want of all things, when the army should be drawn off. And frequently it came to that point, as so large a body of cavalry had been sent abroad in all directions, that the prisoners declared Ambiorix had just then been seen by them in flight, and had not even passed out of sight, so that the hope of overtaking him being raised, and unbounded exertions having been resorted to, those who thought they should acquire the highest favour with Caesar, nearly overcame nature by their ardour, and continually a little only seemed wanting to complete success; but he rescued himself by [means of] lurking-places and forests, and, concealed by the night, made for other districts and quarters, with no greater guard than that of four horsemen, to whom alone he ventured to confide his life.

XLIV. — Having devastated the country in such a manner, Caesar leads back his army with the loss of two cohorts to Durocortorum of the Remi, and, having summoned a council of Gaul to assemble at that place, he resolved to hold an investigation respecting the conspiracy of the Senones and Carnutes, and having pronounced a most severe sentence upon Acco, who had been the contriver of that plot, he punished him after the custom of our ancestors. Some fearing a trial, fled; when he had forbidden these fire and water, he stationed in winter-quarters two legions at the frontiers of the Treviri, two among the Lingones, the remaining six at Agendicum, in the territories of the Senones; and, having provided corn for the army, he set out for Italy, as he had determined, to hold the assizes.

Book VII

I. — Gaul being tranquil, Caesar, as he had determined, sets out for Italy to hold the provincial assizes. There he receives intelligence of the death of Clodius; and, being informed of the decree of the senate [to the effect] that all the youth of Italy should take the military oath, he determined to hold a levy throughout the entire province. Report of these events is rapidly borne into Transalpine Gaul. The Gauls themselves add to the report, and invent what the case seemed to require, [namely] that Caesar was detained by commotions in the city, and could not, amidst so violent dissensions, come to his army. Animated by this opportunity, they who already, previously to this occurrence, were indignant that they were reduced beneath the dominion of Rome, begin to organize their plans for war more openly and daringly. The leading men of Gaul, having convened councils among themselves in the woods, and retired places, complain of the death of Acco: they point out that this fate may fall in turn on themselves: they bewail the unhappy fate of Gaul; and by every sort of promises and rewards, they earnestly solicit some to begin the war, and assert the freedom of Gaul at the hazard of their lives. They say that special care should be paid to this, that Caesar should be cut off from his army, before their secret plans should be divulged. That this was easy, because neither would the legions, in the absence of their general, dare to leave their winter-quarters, nor could the general reach his army without a guard: finally, that it was better to be slain in battle than not to recover their ancient glory in war, and that freedom which they had received from their forefathers.

II. — Whilst these things are in agitation, the Carnutes declare "that they would decline no danger for the sake of the general safety," and promise that they would be the first of all to begin the war; and since they cannot at present take precautions, by giving and receiving hostages, that the affair shall not be divulged they require that a solemn assurance be given them by oath and plighted honour, their military standards being brought together (in which manner their most sacred obligations are made binding), that they should not be deserted by the rest of the Gauls on commencing the war.

III. — When the appointed day came, the Carnutes, under the command of Cotuatus and Conetodunus, desperate men, meet together at Genabum, and slay the Roman citizens who had settled there for the purpose of trading (among the rest, Caius Fusius Cita, a distinguished Roman knight, who by Caesar's orders had presided over the provision department), and plunder their property. The report is quickly spread among all the states of Gaul; for, whenever a more important and remarkable event takes place, they transmit the intelligence through their lands and districts by a shout; the others take it up in succession, and pass it to their neighbours, as happened on this occasion; for the things which were done at Genabum at sunrise were heard in the territories of the Arverni before the end of the first watch, which is an extent of more than a hundred and sixty miles.

IV. — There in like manner, Vercingetorix the son of Celtillus the Arvernian, a young man of the highest power (whose father had held the supremacy of entire Gaul, and had been put to death by his fellow citizens, for this reason, because he aimed at sovereign power), summoned together his dependents, and easily excited them. On his design being made known, they rush to arms: he is expelled from the town of Gergovia by his uncle Gobanitio and the rest of the nobles, who were of opinion, that such an enterprise ought not to be hazarded: he did not however desist, but held in the country a levy of the needy and desperate. Having collected such a body of troops, he brings over to his 30 sentiments such of his fellow citizens as he has access to: he exhorts them to take up arms in behalf of the general freedom, and having assembled great forces he drives from the state his opponents, by whom he had been expelled a short time previously. He is saluted king by his partisans; he sends ambassadors in every direction, he conjures them to adhere firmly to their promise. He quickly attaches to his interests the Senones, Parisii, Pictones, Cadurci, Turones, Aulerci, Lemovice, and all the others who border on the ocean; the supreme command is conferred on him by unanimous consent. On obtaining this authority, he demands hostages from all these states, he orders a fixed number of soldiers to be sent to him immediately; he determines what quantity of arms each state shall prepare at home, and before what time; he pays particular attention to the cavalry. To the utmost vigilance he adds the utmost rigour of authority; and by the severity of his punishments brings over the wavering: for on the commission of a greater crime he puts the perpetrators to death by fire and every sort of tortures; for a slighter cause, he sends home the offenders with their ears cut off, or one of their eyes put out, that they may be an example to the rest, and frighten others by the severity of their punishment.

V.— Having quickly collected an army by their punishments, he sends Lucterius, one of the Cadurci, a man of the utmost daring, with part of his forces, into the territory of the Ruteni; and marches in person into the country of the Bituriges. On his arrival, the Bituriges send ambassadors to the Aedui, under whose protection they were, to solicit aid in order that they might more easily resist the forces of the enemy. The Aedui, by the advice of the lieutenants whom Caesar had left with the army, send supplies of horse and foot to succour the Bituriges. When they came to the river Loire, which separates the Bituriges from the Aedui, they delayed a few days there, and, not daring to pass the river, return home, and send back word to the lieutenants that they had returned through fear of the treachery of the Bituriges, who, they ascertained, had formed this design, that if the Aedui should cross the river, the Bituriges on the one side, and the Arverni on the other, should surround them. Whether they did this for the reason which they alleged to the lieutenants, or influenced by treachery, we think that we ought not to state as certain, because we have no proof. On their departure, the Bituriges immediately unite themselves to the Arverni.

VI. — These affairs being announced to Caesar in Italy at the time when he understood that matters in the city had been reduced to a more tranquil state by the energy of Cneius Pompey, he set out for Transalpine Gaul. After he had arrived there, he was greatly at a loss to know by what means he could reach his army. For if he should summon the legions into the province, he was aware that on their march they would have to fight in his absence; he foresaw too, that if he himself should endeavour to reach the army, he would act injudiciously, in trusting his safety even to those who seemed to be tranquillized.

VII. — In the meantime Lucterius the Cadurcan, having been sent into the country of the Ruteni, gains over that state to the Arverni. Having advanced into the country of the Nitiobriges, and Gabali, he receives hostages from both nations, and, assembling a numerous force, marches to make a descent on the province in the direction of Narbo. Caesar, when this circumstance was announced to him, thought that the march to Narbo ought to take the precedence of all his other plans. When he arrived there, he encourages the timid, and stations garrisons among the Ruteni, in the province of the Volcae Arecomici, and the country around Narbo which was in the vicinity of the enemy; he orders a portion of the forces from the province, and the recruits which he had brought from Italy, to rendezvous among the Helvii who border on the territories of the Arverni.

VIII. — These matters being arranged, and Lucterius now checked and forced to retreat, because he thought it dangerous to enter the line of Roman garrisons, Caesar marches into the country of the Helvii; although mount Cevennes, which separates the Arverni from the Helvii, blocked up the way with very deep snow, as it was the severest season of the year; yet having cleared away the snow to the depth of six feet, and having opened the roads, he reaches the territories of the Arverni, with infinite labour to his soldiers. This people being surprised, because they considered themselves defended by the Cevennes as by a wall, and the paths at this season of the year had never before been passable even to individuals, he orders the cavalry to extend themselves as far as they could, and strike as great a panic as possible into the enemy. These proceedings are speedily announced to Vercingetorix by rumour and his messengers.

Around him all the Arverni crowd in alarm, and solemnly entreat him to protect their property, and not to suffer them to be plundered by the enemy, especially as he saw that all the war was transferred into their country. Being prevailed upon by their entreaties he moves his camp from the country of the Bituriges in the direction of the Arverni.

IX. — Caesar, having delayed two days in that place, because he had anticipated that, in the natural course of events, such would be the conduct of Vercingetorix, leaves the army under pretence of raising recruits and cavalry: he places Brutus, a young man, in command of these forces; he gives him instructions that the cavalry should range as extensively as possible in all directions; that he would exert himself not to be absent from the camp longer than three days. Having arranged these matters, he marches to Vienna by as long journeys as he can, when his own soldiers did not expect him. Finding there a fresh body of cavalry, which he had sent on to that place several days before, marching incessantly night and day, he advanced rapidly through the territory of the Aedui into that of the Lingones, in which two legions were wintering, that, if any plan affecting his own safety should have been organised by the Aedui, he might defeat it by the rapidity of his movements. When he arrived there, he sends information to the rest of the legions, and gathers all his army into one place before intelligence of his arrival could be announced to the Arverni. Vercingetorix, on hearing this circumstance, leads back his army into the country of the Bituriges; and after marching from it to Gergovia, a town of the Boii, whom Caesar had settled there after defeating them in the Helvetian war, and had rendered tributary to the Aedui, he determined to attack it.

X. — This action caused great perplexity to Caesar in the selection of his plans; [he feared] lest, if he should confine his legions in one place for the remaining portion of the winter, all Gaul should revolt when the tributaries of the Aedui were subdued, because it would appear that there was in him no protection for his friends; but if he should draw them too soon out of their winter-quarters, he might be distressed by the want of provisions, in consequence of the difficulty of conveyance. It seemed better, however, to endure every hardship than to alienate the affections of all his allies, by submitting to such an insult. Having, therefore, impressed on the Aedui the necessity of supplying him with provisions, he sends forward messengers to the Boii to inform them of his arrival, and encourage them to remain firm in their allegiance, and resist the attack of the enemy with great resolution. Having left two legions and the luggage of the entire army at Agendicum, he marches to the Boii.

XI. — On the second day, when he came to Vellaunodunum, a town of the Senones, he determined to attack it, in order that he might not leave an enemy in his rear, and might the more easily procure supplies of provisions, and drew a line of circumvallation around it in two days: on the third day, ambassadors being sent from the town to treat of a capitulation, he orders their arms to be brought together, their cattle to be brought forth, and six hundred hostages to be given. He leaves Caius Trebonius, his lieutenant, to complete these arrangements; he himself sets out with the intention of marching as soon as possible to Genabum, a town of the Carnutes, who having then for the first time received information of the siege of Vellaunodunum, as they thought that it would be protracted to a longer time, were preparing a garrison to send to Genabum for the defence of that town. Caesar arrived here in two days; after pitching his camp before the town, being prevented by the time of the day, he defers the attack to the next day, and orders his soldiers to prepare whatever was necessary for that enterprise; and as a bridge over the Loire connected the town of Genabum with the opposite bank, fearing lest the inhabitants should escape by night from the town, he orders two legions to keep watch under arms. The people of Genabum came forth silently from the city before midnight, and began to cross the river. When this circumstance was announced by scouts, Caesar, having set fire to the gates, sends in the legions which he had ordered to be ready, and obtains possession of the town so completely, that very few of the whole number of the enemy escaped being taken alive, because the narrowness of the bridge and the roads prevented the multitude from escaping. He pillages and burns the town, gives the booty to the soldiers, then leads his army over the Loire, and marches into the territories of the Bituriges.

XII. — Vercingetorix, when he ascertained the arrival of Caesar, desisted from the siege [of Gergovia], and marched to meet Caesar. The latter had commenced to besiege Noviodunum; and when ambassadors came from this town to beg that he would pardon them and spare their lives, in order that he might execute the rest of his designs with the rapidity by which he had accomplished most of them, he orders their arms to be collected, their horses to be brought forth, and hostages to be given. A part of the hostages being now delivered up, when the rest of the terms were being performed, a few centurions and soldiers being sent into the town to collect the arms and horses, the enemy's cavalry, which had outstripped the main body of Vercingetorix's army, was seen at a distance; as soon as the townsmen beheld them, and entertained hopes of assistance, raising a shout, they began to take up arms, shut the gates, and line the walls. When the centurions in the town understood from the signal-making of the Gauls that they were forming some new design, they drew their swords and seized the gates, and recovered all their men safe.

XIII. — Caesar orders the horse to be drawn out of the camp, and commences a cavalry action. His men being now distressed, Caesar sends to their aid about four hundred German horse, which he had determined, at the beginning, to keep with himself. The Gauls could not withstand their attack, but were put to flight, and retreated to their main body, after losing a great number of men. When they were routed, the townsmen, again intimidated, arrested those persons by whose exertions they thought that the mob had been roused, and brought them to Caesar, and surrendered themselves to him. When these affairs were accomplished, Caesar marched to the Avaricum, which was the largest and best fortified town in the territories of the Bituriges, and situated in a most fertile tract of country; because he confidently expected that on taking that town, he would reduce beneath his dominion the state of the Bituriges.

XIV. — Vercingetorix, after sustaining such a series of losses at Vellaunodunum, Genabum, and Noviodunum, summons his men to a council. He impresses on them "that the war must be prosecuted on a very different system from that which had been previously adopted; but they should by all means aim at this object, that the Romans should be prevented from foraging and procuring provisions; that this was easy, because they themselves were well supplied with cavalry and were likewise assisted by the season of the year; that forage could not be cut; that the enemy must necessarily disperse, and look for it in the houses, that all these might be daily destroyed by the horse. Besides that the interests of private property must be neglected for the sake of the general safety; that the villages and houses ought to be fired, over such an extent of country in every direction from Boia, as the Romans appeared capable of scouring in their search for forage. That an abundance of these necessaries could be supplied to them, because they would be assisted by the resources of those in whose territories the war would be waged: that the Romans either would not bear the privation, or else would advance to any distance from the camp with considerable danger; and that it made no difference whether they slew them or stripped them of their baggage, since, if it was lost, they could not carry on the war. Besides that, the towns ought to be burnt which were not secured against every danger by their fortifications or natural advantages; that there should not be places of retreat for their own countrymen for declining military service, nor be exposed to the Romans as inducements to carry off abundance of provisions and plunder. If these sacrifices should appear heavy or galling, that they ought to consider it much more distressing that their wives and children should be dragged off to slavery, and themselves slain; the evils which must necessarily befall the conquered.

XV. — This opinion having been approved of by unanimous consent, more than twenty towns of the Bituriges are burnt in one day. Conflagrations are beheld in every quarter; and although all bore this with great regret, yet they laid before themselves this consolation, that, as the victory was certain, they could quickly recover their losses. There is a debate concerning Avaricum in the general council, whether they should decide that it should be burnt or defended. The Bituriges threw themselves at the feet of all the Gauls, and entreat that they should not be compelled to set fire with their own hands to the fairest city of almost the whole of Gaul, which was both a protection and ornament to the state; they say that "they could easily defend it, owing to the nature of the ground, for, being enclosed almost on every side by a river and a marsh, it had only one entrance, and that very narrow." Permission being granted to them at their earnest request, Vercingetorix at first dissuades them from it, but afterwards concedes the point, owing to their entreaties and the compassion of the soldiers. A proper garrison is selected for the town.

XVI. — Vercingetorix follows closely upon Caesar by shorter marches, and selects for his camp a place defended by woods and marshes, at the distance of fifteen miles from Avaricum. There he received intelligence by trusty scouts, every hour in the day, of what was going on at Avaricum, and ordered whatever he wished to be done; he closely watched all our expeditions for corn and forage, and whenever they were compelled to go to a greater distance, he attacked them when dispersed, and inflicted severe loss upon them; although the evil was remedied by our men, as far as precautions could be taken, by going forth at irregular times, and by different ways.

XVII. — Caesar pitching his camp at that side of the town which was not defended by the river and marsh, and had a very narrow approach, as we have mentioned, began to raise the vineae and erect two towers; for the nature of the place prevented him from drawing a line of circumvallation. He never ceased to importune the Boii and Aedui for supplies of corn; of whom the one [the Aedui], because they were acting with no zeal, did not aid him much; the others [the Boii], as their resources were not great, quickly consumed what they had. Although the army was distressed by the greatest want of corn, through the poverty of the Boii, the apathy of the Aedui, and the burning of the houses, to such a degree, that for several days the soldiers were without corn, and satisfied their extreme hunger with cattle driven from the remote villages; yet no language was heard from them unworthy of the majesty of the Roman people and their former victories. Moreover, when Caesar addressed the legions, one by one, when at work, and said that he would raise the siege, if they felt the scarcity too severely, they unanimously begged him "not to do so; that they had served for several years under his command in such a manner, that they never submitted to insult, and never abandoned an enterprise without accomplishing it; that they should consider it a disgrace if they abandoned the siege after commencing it; that it was better to endure every hardship than not to avenge the manes of the Roman citizens who perished at Genabum by the perfidy of the Gauls." They entrusted the same declarations to the centurions and military tribunes, that through them they might be communicated to Caesar.

XVIII. — When the towers had now approached the walls, Caesar ascertained from the captives that Vercingetorix, after destroying the forage, had pitched his camp nearer Avaricum, and that he himself with the cavalry and light-armed infantry, who generally fought among the horse, had gone to lay an ambuscade in that quarter to which he thought that our troops would come the next day to forage. On learning these facts, he set out from the camp secretly at midnight, and reached the camp of the enemy early in the morning. They having quickly learned the arrival of Caesar by scouts, hid their cars and baggage in the thickest parts of the woods, and drew up all their forces in a lofty and open space: which circumstance being announced, Caesar immediately ordered the baggage to be piled, and the arms to be got ready.

XIX. — There was a hill of a gentle ascent from the bottom; a dangerous and impassable marsh, not more than fifty feet broad, begirt it on almost every side. The Gauls, having broken down the bridges, posted themselves on this hill, in confidence of their position, and being drawn up in tribes according to their respective states, held all the fords and passages of that marsh with trusty guards, thus determined that if the Romans should attempt to force the marsh, they would overpower them from the higher ground while sticking in it, so that whoever saw the nearness of the position, would imagine that the two armies were prepared to fight on almost equal terms; but whoever should view accurately the disadvantage of position, would discover that they were showing off an empty affectation of courage. Caesar clearly points out to his soldiers, who were indignant that the enemy could bear the sight of them at the distance of so short a space, and were earnestly demanding the signal for action, "with how great loss and the death of how many gallant men the victory would necessarily be purchased: and when he saw them so determined to decline no danger for his renown, that he ought to be considered guilty of the utmost injustice if he did not hold their life dearer than his own personal safety." Having thus consoled his soldiers, he leads them back on the same day to the camp, and determined to prepare the other things which were necessary for the siege of the town.

XX. — Vercingetorix, when he had returned to his men, was accused of treason, in that he had moved his camp nearer the Romans, in that he had gone away with all the cavalry, in that he had left so great forces without a commander, in that, on his departure, the Romans had come at such a favourable season, and with such despatch; that all these circumstances could not have happened accidentally or without design; that he preferred holding the sovereignty of Gaul by the grant of Caesar, to acquiring it by their favour. Being accused in such a manner, he made the following reply to these charges: — "That his moving his camp had been caused by want of forage, and had been done even by their advice; that his approaching near the Romans had been a measure dictated by the favourable nature of the ground, which would defend him by its natural strength; that the service of the cavalry could not have been requisite in marshy ground, and was useful in that place to which they had gone; that he, on his departure, had given the supreme command to no one intentionally, lest he should be induced by the eagerness of the multitude to hazard an engagement, to which he perceived that all were inclined, owing to their want of energy, because they were unable to endure fatigue any longer. That, if the Romans in the meantime came up by chance, they [the Gauls] should feel grateful to fortune; if invited by the information of some one they should feel grateful to him, because they were enabled to see distinctly from the higher ground the smallness of the number of their enemy, and despise the courage of those who, not daring to fight, retreated disgracefully into their camp. That he desired no power from Caesar by treachery, since he could have it by victory, which was now assured to himself and to all the Gauls; nay, that he would even give them back the command, if they thought that they conferred honour on him, rather then received safety from him. That you may be assured," said he, "that I speak these words with truth; — listen to these Roman soldiers!" He produces some camp-followers whom he had surprised on a foraging expedition some days before, and had tortured by famine and confinement. They being previously instructed in what answers they should make when examined, say, "That they were legionary soldiers, that, urged by famine and want, they had recently gone forth from the camp, [to see] if they could find any corn or cattle in the fields; that the whole army was distressed by a similar scarcity, nor had any one now sufficient strength, nor could bear the labour of the work; and therefore that the general was determined, if he made no progress in the siege, to draw off his army in three days." "These benefits," says Vercingetorix, "you receive from me, whom you accuse of treason — me, by whose exertions you see so powerful and victorious an army almost destroyed by famine, without shedding one drop of your blood; and I have taken precautions that no state shall admit within its territories this army in its ignominious flight from this place."

XXI. — The whole multitude raise a shout and clash their arms, according to their custom, as they usually do in the case of him whose speech they approve; [they exclaim] that Vercingetorix was a consummate general, and that they had no doubt of his honour; that the war could not be conducted with greater prudence. They determine that ten thousand men should be picked out of the entire army and sent into the town, and decide that the general safety should not be entrusted to the Bituriges alone, because they were aware that the glory of the victory must rest with the Bituriges, if they made good the defence of the town.

XXII. — To the extraordinary valour of our soldiers, devices of every sort were opposed by the Gauls; since they are a nation of consummate ingenuity, and most skilful in imitating and making those things which are imparted by any one; for they turned aside the hooks with nooses, and when they had caught hold of them firmly, drew them on by means of engines, and undermined the mound the more skilfully on this account, because there are in their territories extensive iron mines, and consequently every description of mining operations is known and practised by them. They had furnished, moreover, the whole wall on every side with turrets, and had covered them with skins. Besides, in their frequent sallies by day and night, they attempted either to set fire to the mound, or attack our soldiers when engaged in the works; and, moreover, by splicing the upright timbers of their own towers, they equalled the height of ours, as fast as the mound had daily raised them, and countermined our mines, and impeded the working of them by stakes bent and sharpened at the ends, and boiling pitch, and stones of very great weight, and prevented them from approaching the walls.

XXIII. — But this is usually the form of all the Gallic walls. Straight beams, connected lengthwise and two feet distant from each other at equal intervals, are placed together on the ground; these are mortised on the inside, and covered with plenty of earth. But the intervals which we have mentioned, are closed up in front by large stones. These being thus laid and cemented together, another row is added above, in such a manner that the same interval may be observed, and that the beams may not touch one another, but equal spaces intervening, each row of beams is kept firmly in its place by a row of stones. In this manner the whole wall is consolidated, until the regular height of the wall be completed. This work, with respect to appearance and variety, is not unsightly, owing to the alternate rows of beams and stones, which preserve their order in right lines; and, besides, it possesses great advantages as regards utility and the defence of cities; for the stone protects it from fire, and the wood from the battering ram, since it [the wood] being mortised in the inside with rows of beams, generally forty feet each in length, can neither be broken through nor torn asunder.

XXIV. — The siege having been impeded by so many disadvantages, the soldiers, although they were retarded during the whole time, by the mud, cold, and constant showers, yet by their incessant labour overcame all these obstacles, and in twenty-five days raised a mound three hundred and thirty feet broad and eighty feet high. When it almost touched the enemy's walls, and Caesar, according to his usual custom, kept watch at the work, and encouraged the soldiers not to discontinue the work for a moment: a little before the third watch they discovered that the mound was sinking, since the enemy had set it on fire by a mine; and at the same time a shout was raised along the entire wall, and a sally was made from two gates on each side of the turrets. Some at a distance were casting torches and dry wood from the wall on the mound, others were pouring on it pitch, and other materials, by which the flame might be excited, so that a plan could hardly be formed, as to where they should first run to the defence, or to what part aid should be brought. However, as two legions always kept guard before the camp by Caesar's orders, and several of them were at stated times at the work, measures were promptly taken, that some should oppose the sallying party, others draw back the towers and make a cut in the rampart; and moreover, that the whole army should hasten from the camp to extinguish the flames.

XXV. — When the battle was going on in every direction, the rest of the night being now spent, and fresh hopes of victory always arose before the enemy: the more so on this account because they saw the coverings of our towers burnt away, and perceived that we, being exposed, could not easily go to give assistance, and they themselves were always relieving the weary with fresh men, and considered that all the safety of Gaul rested on this crisis; there happened in my own view a circumstance which, having appeared to be worthy of record, we thought it ought not to be omitted. A certain Gaul before the gate of the town, who was casting into the fire opposite the turret balls of tallow and fire which were passed along to him, was pierced with a dart on the right side and fell dead. One of those next him stepped over him as he lay, and discharged the same office: when the second man was slain in the same manner by a wound from a cross-bow, a third succeeded him, and a fourth succeeded the third: nor was this post left vacant by the besieged, until, the fire of the mound having been extinguished, and the enemy repulsed in every direction, an end was put to the fighting.

XXVI. — The Gauls having tried every expedient, as nothing had succeeded, adopted the design of fleeing from the town the next day, by the advice and order of Vercingetorix. They hoped that, by attempting it at the dead of night, they would effect it without any great loss of men, because the camp of Vercingetorix was not far distant from the town, and the extensive marsh which intervened was likely to retard the Romans in the pursuit. And they were now preparing to execute this by night, when the matrons suddenly ran out into the streets, and weeping cast themselves at the feet of their husbands, and requested of them, with every entreaty, that they should not abandon themselves and their common children to the enemy for punishment, because the weakness of their nature and physical powers prevented them from taking to flight. When they saw that they (as fear does not generally admit of mercy in extreme danger) persisted in their resolution, they began to shout aloud, and give intelligence of their flight to the Romans. The Gauls being intimidated by fear of this, lest the passes should be pre-occupied by the Roman cavalry, desisted from their design.

XXVII. — The next day Caesar, the tower being advanced, and the works which he had determined to raise being arranged, a violent storm arising, thought this no bad time for executing his designs, because he observed the guards arranged on the walls a little too negligently, and therefore ordered his own men to engage in their work more remissly, and pointed out what he wished to be done. He drew up his soldiers in a secret position within the vineae, and exhorts them to reap, at least, the harvest of victory proportionate to their exertions. He proposed a reward for those who should first scale the walls, and gave the signal to the soldiers. They suddenly flew out from all quarters and quickly filled the wall.

XXVIII. — The enemy being alarmed by the suddenness of the attack, were dislodged from the wall and towers, and drew up, in form of a wedge, in the market-place and the open streets, with this intention that, if an attack should be made on any side, they should fight with their line drawn up to receive it. When they saw no one descending to the level ground, and the enemy extending themselves along the entire wall in every direction, fearing lest every hope of flight should be cut off, they cast away their arms, and sought, without stopping, the most remote parts of the town. A part was then slain by the infantry when they were crowding upon one another in the narrow passage of the gates; and a part having got without the gates, were cut to pieces by the cavalry: nor was there one who was anxious for the plunder. Thus, being excited by the massacre at Genabum and the fatigue of the siege, they spared neither those worn out with years, women, or children. Finally, out of all that number, which amounted to about forty thousand, scarcely eight hundred, who fled from the town when they heard the first alarm, reached Vercingetorix in safety: and he, the night being now far spent, received them in silence after their flight (fearing that any sedition should arise in the camp from their entrance in a body and the compassion of the soldiers), so that, having arranged his friends and the chiefs of the states at a distance on the road, he took precautions that they should be separated and conducted to their fellow countrymen, to whatever part of the camp had been assigned to each state from the beginning.

XXIX. — Vercingetorix having convened an assembly on the following day, consoled and encouraged his soldiers in the following words: — "That they should not be too much depressed in spirit, nor alarmed at their loss; that the Romans did not conquer by valour nor in the field, but by a kind of art and skill in assault, with which they themselves were unacquainted; that whoever expected every event in the war to be favourable, erred; that it never was his opinion that Avaricum should be defended, of the truth of which statement he had themselves as witnesses, but that it was owing to the imprudence of the Bituriges, and the too ready compliance of the rest, that this loss was sustained; that, however, he would soon compensate it by superior advantages; for that he would, by his exertions, bring over those states which severed themselves from the rest of the Gauls, and would create a general unanimity throughout the whole of Gaul, the union of which not even the whole earth could withstand, and that he had it already almost effected; that in the meantime it was reasonable that he should prevail on them, for the sake of the general safety, to begin to fortify their camp, in order that they might the more easily sustain the sudden attacks of the enemy."

XXX. — This speech was not disagreeable to the Gauls, principally, because he himself was not disheartened by receiving so severe a loss, and had not concealed himself, nor shunned the eyes of the people: and he was believed to possess greater foresight and sounder judgment than the rest, because, when the affair was undecided, he had at first been of opinion that Avaricum should be burnt, and afterwards that it should be abandoned. Accordingly, as ill success weakens the authority of other generals, so, on the contrary, his dignity increased daily, although a loss was sustained: at the same time they began to entertain hopes, on his assertion, of uniting the rest of the states to themselves, and on this occasion, for the first time, the Gauls began to fortify their camps, and were so alarmed that although they were men unaccustomed to toil, yet they were of opinion that they ought to endure and suffer everything which should be imposed upon them.

XXXI. — Nor did Vercingetorix use less efforts than he had promised, to gain over the other states, and [in consequence] endeavoured to entice their leaders by gifts and promises. For this object he selected fitting emissaries by whose subtle pleading or private friendship each of the nobles could be most easily influenced. He takes care that those who fled to him on the storming of Avaricum should be provided with arms and clothes. At the same time, that his diminished forces should be recruited, he levies a fixed quota of soldiers from each state, and defines the number and day before which he should wish them brought to the camp, and orders all the archers, of whom there was a very great number in Gaul, to be collected and sent to him. By these means, the troops which were lost at Avaricum are speedily replaced. In the meantime, Teutomarus, the son of Ollovicon, the king of the Nitiobriges, whose father had received the appellation of friend from our senate, came to him with a great number of his own horse and those whom he had hired from Aquitania.

XXXII. — Caesar, after delaying several days at Avaricum, and finding there the greatest plenty of corn and other provisions, refreshed his army after their fatigue and privation. The winter being almost ended, when he was invited by the favourable season of the year to prosecute the war and march against the enemy, [and try] whether he could draw them from the marshes and woods, or else press them by a blockade; some noblemen of the Aedui came to him as ambassadors to entreat "that in an extreme emergency he should succour their state; that their affairs were in the utmost danger, because, whereas single magistrates had been usually appointed in ancient times and held the power of king for a single year, two persons now exercised this office, and each asserted that he was appointed according to their laws. That one of them was Convictolitanis, a powerful and illustrious youth; the other Cotus, sprung from a most ancient family, and personally a man of very great influence and extensive connections. His brother Valetiacus had borne the same office during the last year: that the whole state was up in arms; the senate divided, the people divided; that each of them had his own adherents; and that, if the animosity would be fomented any longer the result would be that one part of the state would come to a collision with the other; that it rested with his activity and influence to prevent it."

XXXIII. — Although Caesar considered it ruinous to leave the war and the enemy, yet, being well aware what great evils generally arise from internal dissensions, lest a state so powerful and so closely connected with the Roman people, which he himself had always fostered and honoured in every respect, should have recourse to violence and arms, and that the party which had less confidence in its own power should summon aid from Vercingetorix, he determined to anticipate this movement; and because, by the laws of the Aedui, it was not permitted those who held the supreme authority to leave the country, he determined to go in person to the Aedui, lest he should appear to infringe upon their government and laws, and summoned all the senate, and those between whom the dispute was, to meet him at Decetia. When almost all the state had assembled there, and he was informed that one brother had been declared magistrate by the other, when only a few persons were privately summoned for the purpose, at a different time and place from what he ought, whereas the laws not only forbade two belonging to one family to be elected magistrates while each was alive, but even deterred them from being in the senate, he compelled Cotus to resign his office; he ordered Convictolitanis, who had been elected by the priests, according to the usage of the state, in the presence of the magistrates, to hold the supreme authority.

XXXIV. — Having pronounced this decree between [the contending parties], he exhorted the Aedui to bury in oblivion their disputes and dissensions, and, laying aside all these things, devote themselves to the war, and expect from him, on the conquest of Gaul, those rewards which they should have earned, and send speedily to him all their cavalry and ten thousand infantry, which he might place in different garrisons to protect his convoys of provisions, and then divided his army into two parts: he gave Labienus four legions to lead into the country of the Senones and Parisii; and led in person six into the country of the Arverni, in the direction of the town of Gergovia, along the banks of the Allier. He gave part of the cavalry to Labienus, and kept part to himself. Vercingetorix, on learning this circumstance, broke down all the bridges over the river and began to march on the other bank of the Allier.

XXXV. — When each army was in sight of the other, and was pitching their camp almost opposite that of the enemy, scouts being distributed in every quarter, lest the Romans should build a bridge and bring over their troops; it was to Caesar a matter attended with great difficulties, lest he should be hindered from passing the river during the greater part of the summer, as the Allier cannot generally be forded before the autumn. Therefore, that this might not happen, having pitched his camp in a woody place opposite to one of those bridges which Vercingetorix had taken care should be broken down, the next day he stopped behind with two legions in a secret place: he sent on the rest of the forces as usual, with all the baggage, after having selected some cohorts, that the number of the legions might appear to be complete. Having ordered these to advance as far as they could, when now, from the time of day, he conjectured they had come to an encampment, he began to rebuild the bridge on the same piles, the lower part of which remained entire. Having quickly finished the work and led his legions across, he selected a fit place for a camp, and recalled the rest of his troops. Vercingetorix, on ascertaining this fact, went before him by forced marches, in order that he might not be compelled to come to an action against his will.

XXXVI. — Caesar, in five days' march, went from that place to Gergovia, and after engaging in a slight cavalry skirmish that day, on viewing the situation of the city, which, being built on a very high mountain, was very difficult of access, he despaired of taking it by storm, and determined to take no measures with regard to besieging it before he should secure a supply of provisions. But Vercingetorix, having pitched his camp on the mountain near the town, placed the forces of each state separately and at small intervals around himself, and having occupied all the hills of that range as far as they commanded a view [of the Roman encampment], he presented a formidable appearance; he ordered the rulers of the states, whom he had selected as his council of war, to come to him daily at the dawn, whether any measure seemed to require deliberation or execution. Nor did he allow almost any day to pass without testing in a cavalry action, the archers being intermixed, what spirit and valour there was in each of his own men. There was a hill opposite the town, at the very foot of that mountain, strongly fortified and precipitous on every side (which if our men could gain, they seemed likely to exclude the enemy from a great share of their supply of water, and from free foraging; but this place was occupied by them with a weak garrison): however, Caesar set out from the camp in the silence of night, and dislodging the garrison before succour could come from the town, he got possession of the place and posted two legions there, and drew from the greater camp to the less a double trench twelve feet broad, so that the soldiers could even singly pass secure from any sudden attack of the enemy.

XXXVII. — Whilst these affairs were going on at Gergovia, Convictolitanis, the Aeduan, to whom we have observed the magistracy was adjudged by Caesar, being bribed by the Arverni, holds a conference with certain young men, the chief of whom were Litavicus and his brothers, who were born of a most noble family. He shares the bribe with them, and exhorts them to "remember that they were free and born for empire; that the state of the Aedui was the only one which retarded the most certain victory of the Gauls; that the rest were held in check by its authority; and, if it was brought over, the Romans would not have room to stand on in Gaul; that he had received some kindness from Caesar, only so far, however, as gaining a most just cause by his decision; but that he assigned more weight to the general freedom; for, why should the Aedui go to Caesar to decide concerning their rights and laws, rather than the Romans come to the Aedui?" The young men being easily won over by the speech of the magistrate and the bribe, when they declared that they would even be leaders in the plot, a plan for accomplishing it was considered, because they were confident their state could not be induced to undertake the war on slight grounds. It was resolved that Litavicus should have the command of the ten thousand which were being sent to Caesar for the war, and should have charge of them on their march, and that his brothers should go before him to Caesar. They arrange the other measures, and the manner in which they should have them done.

XXXVIII. — Litavicus, having received the command of the army, suddenly convened the soldiers, when he was about thirty miles distant from Gergovia, and, weeping, said, "Soldiers, whither are we going? All our knights and all our nobles have perished. Eporedorix and Viridomarus, the principal men of the state, being accused of treason, have been slain by the Romans without even permission to plead their cause. Learn this intelligence from those who have escaped from the massacre; for I, since my brothers and all my relations have been slain, am prevented by grief from declaring what has taken place." Persons are brought forward whom he had instructed in what he would have them say, and make the same statements to the soldiery as Litavicus had made: that all the knights of the Aedui were slain because they were said to have held conferences with the Arverni; that they had concealed themselves among the multitude of soldiers, and had escaped from the midst of the slaughter. The Aedui shout aloud and conjure Litavicus to provide for their safety. "As if," said he, "it were a matter of deliberation, and not of necessity, for us to go to Gergovia and unite ourselves to the Arverni. Or have we any reasons to doubt that the Romans, after perpetrating the atrocious crime, are now hastening to slay us? Therefore, if there be any spirit in us, let us avenge the death of those who have perished in a most unworthy manner, and let us slay these robbers." He points to the Roman citizens, who had accompanied them, in reliance on his protection. He immediately seizes a great quantity of corn and provisions, cruelly tortures them, and then puts them to death, sends messengers throughout the entire state of the Aedui, and rouses them completely by the same falsehood concerning the slaughter of their knights and nobles; he earnestly advises them to avenge, in the same manner as he did, the wrongs which they had received.

XXXIX. — Eporedorix, the Aeduan, a young man born in the highest rank and possessing very great influence at home, and, along with Viridomarus, of equal age and influence, but of inferior birth, whom Caesar had raised from a humble position to the highest rank, on being recommended to him by Divitiacus, had come in the number of horse, being summoned by Caesar by name. These had a dispute with each other for precedence, and in the struggle between the magistrates they had contended with their utmost efforts, the one for Convictolitanis, the other for Cotus. Of these Eporedorix, on learning the design of Litavicus, lays the matter before Caesar almost at midnight; he entreats that Caesar should not suffer their state to swerve from the alliance with the Roman people, owing to the depraved counsels of a few young men, which he foresaw would be the consequence if so many thousand men should unite themselves to the enemy, as their relations could not neglect their safety, nor the state regard it as a matter of slight importance.

XL. — Caesar felt great anxiety on this intelligence, because he had always especially indulged the state of the Aedui, and, without any hesitation, draws out from the camp four light-armed legions and all the cavalry: nor had he time, at such a crisis, to contract the camp, because the affair seemed to depend upon despatch. He leaves Caius Fabius, his lieutenant, with two legions to guard the camp. When he ordered the brothers of Litavicus to be arrested, he discovers that they had fled a short time before to the camp of the enemy. He encouraged his soldiers "not to be disheartened by the labour of the journey on such a necessary occasion," and, after advancing twenty-five miles, all being most eager, he came in sight of the army of the Aedui, and, by sending on his cavalry, retards and impedes their march; he then issues strict orders to all his soldiers to kill no one. He commands Eporedorix and Viridomarus, who they thought were killed, to move among the cavalry and address their friends. When they were recognized and the treachery of Litavicus discovered, the Aedui began to extend their hands to intimate submission, and, laying down their arms, to deprecate death. Litavicus, with his clansmen, who after the custom of the Gauls consider it a crime to desert their patrons, even in extreme misfortune, flees forth to Gergovia.

XLI. — Caesar, after sending messengers to the state of the Aedui, to inform them that they whom he could have put to death by the right of war were spared through his kindness, and after giving three hours of the night to his army for his repose, directed his march to Gergovia. Almost in the middle of the journey, a party of horse that were sent by Fabius stated in how great danger matters were; they inform him that the camp was attacked by a very powerful army, while fresh men were frequently relieving the wearied, and exhausting our soldiers by the incessant toil, since, on account of the size of the camp, they had constantly to remain on the rampart; that many had been wounded by the immense number of arrows and all kinds of missiles; that the engines were of great service in withstanding them; that Fabius, at their departure, leaving only two gates open, was blocking up the rest, and was adding breast-works to the ramparts, and was preparing himself for a similar casualty on the following day. Caesar, after receiving this information, reached the camp before sunrise owing to the very great zeal of his soldiers.

XLII. — Whilst these things are going on at Gergovia, the Aedui, on receiving the first announcements from Litavicus, leave themselves no time to ascertain the truth of these statements. Some are stimulated by avarice, others by revenge and credulity, which is an innate propensity in that race of men to such a degree that they consider a slight rumour as an ascertained fact. They plunder the property of the Roman citizens, and either massacre them or drag them away to slavery. Convictolitanis increases the evil state of affairs, and goads on the people to fury, that by the commission of some outrage they may be ashamed to return to propriety. They entice from the town of Cabillonus, by a promise of safety, Marcus Aristius, a military tribune, who was on his march to his legion; they compel those who had settled there for the purpose of trading to do the same. By constantly attacking them on their march they strip them of all their baggage; they besiege day and night those that resisted; when many were slain on both sides, they excite a greater number to arms.

XLIII. — In the meantime, when intelligence was brought that all their soldiers were in Caesar's power, they run in a body to Aristius; they assure him that nothing had been done by public authority; they order an inquiry to be made about the plundered property; they confiscate the property of Litavicus and his brothers; they send ambassadors to Caesar for the purpose of clearing themselves. They do all this with a view to recover their soldiers; but being contaminated by guilt, and charmed by the gains arising from the plundered property, as that act was shared in by many, and being tempted by the fear of punishment, they began to form plans of war and stir up the other states by embassies. Although Caesar was aware of this proceeding, yet he addresses the ambassadors with as much mildness as he can: "That he did not think worse of the state on account of the ignorance and fickleness of the mob, nor would diminish his regard for the Aedui." He himself, fearing a greater commotion in Gaul, in order to prevent his being surrounded by all the states, began to form plans as to the manner in which he should return from Gergovia and again concentrate his forces, lest a departure arising from the fear of a revolt should seem like a flight.

XLIV. — Whilst he was considering these things an opportunity of acting successfully seemed to offer. For, when he had come into the smaller camp for the purpose of securing the works, he noticed that the hill in the possession of the enemy was stript of men, although, on the former days, it could scarcely be seen on account of the numbers on it. Being astonished, he inquires the reason of it from the deserters, a great number of whom flocked to him daily. They all concurred in asserting, what Caesar himself had already ascertained by his scouts, that the back of that hill was almost level; but likewise woody and narrow, by which there was a pass to the other side of the town; that they had serious apprehensions for this place, and had no other idea, on the occupation of one hill by the Romans, than that, if they should lose the other, they would be almost surrounded, and cut off from all egress and foraging; that they were all summoned by Vercingetorix to fortify this place.

XLV. — Caesar, on being informed of this circumstance, sends several troops of horse to the place immediately after midnight; he orders them to range in every quarter with more tumult than usual. At dawn he orders a large quantity of baggage to be drawn out of the camp, and the muleteers with helmets, in the appearance and guise of horsemen, to ride round the hills. To these he adds a few cavalry, with instructions to range more widely to make a show. He orders them all to seek the same quarter by a long circuit; these proceedings were seen at a distance from the town, as Gergovia commanded a view of the camp, nor could the Gauls ascertain at so great a distance what certainty there was in the manoeuvre. He sends one legion to the same hill, and after it had marched a little, stations it in the lower ground, and conceals it in the woods. The suspicions of the Gauls are increased, and all their forces are marched to that place to defend it. Caesar, having perceived the camp of the enemy deserted, covers the military insignia of his men, conceals the standards, and transfers his soldiers in small bodies from the greater to the less camp, and points out to the lieutenants whom he had placed in command over the respective legions, what he should wish to be done; he particularly advises them to restrain their men from advancing too far, through their desire of fighting, or their hope of plunder; he sets before them what disadvantages the unfavourable nature of the ground carries with it; that they could be assisted by despatch alone: that success depended on a surprise, and not on a battle. After stating these particulars, he gives the signal for action, and detaches the Aedui at the same time by another ascent on the right.

XLVI. — The town wall was 1,200 paces distant from the plain and foot of the ascent, in a straight line, if no gap intervened; whatever circuit was added to this ascent, to make the hill easy, increased the length of the route. But almost in the middle of the hill, the Gauls had previously built a wall six feet high, made of large stones, and extending in length as far as the nature of the ground permitted, as a barrier to retard the advance of our men; and leaving all the lower space empty, they had filled the upper part of the hill, as far as the wall of the town, with their camps very close to one another. The soldiers, on the signal being given, quickly advance to this fortification, and passing over it, make themselves masters of the separate camps. And so great was their activity in taking the camps, that Teutomarus, the king of the Nitiobriges, being suddenly surprised in his tent, as he had gone to rest at noon, with difficulty escaped from the hands of the plunderers, with the upper part of his person naked, and his horse wounded.

XLVII. — Caesar, having accomplished the object which he had in view, ordered the signal to be sounded for a retreat; and the soldiers of the tenth legion, by which he was then accompanied, halted. But the soldiers of the other legions, not hearing the sound of the trumpet, because there was a very large valley between them, were however kept back by the tribunes of the soldiers and the lieutenants, according to Caesar's orders; but being animated by the prospect of speedy victory, and the flight of the enemy, and the favourable battles of former periods, they thought nothing so difficult that their bravery could not accomplish it; nor did they put an end to the pursuit, until they drew nigh to the wall of the town and the gates. But then, when a shout arose in every quarter of the city, those who were at a distance being alarmed by the sudden tumult, fled hastily from the town, since they thought that the enemy were within the gates. The matrons begin to cast their clothes and silver over the wall, and bending over as far as the lower part of the bosom, with outstretched hands beseech the Romans to spare them, and not to sacrifice to their resentment even women and children, as they had done at Avaricum. Some of them let themselves down from the walls by their hands, and surrendered to our soldiers. Lucius Fabius, a centurion of the eighth legion, who, it was ascertained, had said that day among his fellow soldiers that he was excited by the plunder of Avaricum, and would not allow any one to mount the wall before him, finding three men of his own company, and being raised up by them, scaled the wall. He himself, in turn, taking hold of them one by one, drew them up to the wall.

XLVIII. — In the meantime those who had gone to the other part of the town to defend it, as we have mentioned above, at first, aroused by hearing the shouts, and, afterwards, by frequent accounts that the town was in possession of the Romans, sent forward their cavalry, and hastened in larger numbers to that quarter. As each first came he stood beneath the wall, and increased the number of his countrymen engaged in action. When a great multitude of them had assembled, the matrons, who a little before were stretching their hands from the walls to the Romans, began to beseech their countrymen, and after the Gallic fashion to show their dishevelled hair, and bring their children into public view. Neither in position nor in numbers was the contest an equal one to the Romans; at the same time, being exhausted by running and the long continuation of the fight, they could not easily withstand fresh and vigorous troops.

XLIX. — Caesar, when he perceived that his soldiers were fighting on unfavourable ground, and that the enemy's forces were increasing, being alarmed for the safety of his troops, sent orders to Titus Sextius, one of his lieutenants, whom he had left to guard the smaller camp, to lead out his cohorts quickly from the camp, and post them at the foot of the hill, on the right wing of the enemy; that if he should see our men driven from the ground, he should deter the enemy from following too closely. He himself, advancing with the legion a little from that place where he had taken his post, awaited the issue of the battle.

L. — While the fight was going on most vigorously, hand to hand, and the enemy depended on their position and numbers, our men on their bravery, the Aedui suddenly appeared on our exposed flank, as Caesar had sent them by another ascent on the right, for the sake of creating a diversion. These, from the similarity of their arms, greatly terrified our men; and although they were discovered to have their right shoulders bare, which was usually the sign of those reduced to peace, yet the soldiers suspected that this very thing was done by the enemy to deceive them. At the same time Lucius Fabius the centurion, and those who had scaled the wall with him, being surrounded and slain, were cast from the wall. Marcus Petreius, a centurion of the same legion, after attempting to hew down the gates, was overpowered by numbers, and, despairing of his safety, having already received many wounds, said to the soldiers of his own company who followed him: "Since I cannot save you as well as myself, I shall at least provide for your safety, since I allured by the love of glory, led you into this danger, do you save yourselves when an opportunity is given." At the same time he rushed into the midst of the enemy, and slaying two of them, drove back the rest a little from the gate. When his men attempted to aid him, "In vain," he says, "you endeavour to procure my safety since blood and strength are now failing me, therefore leave this, while you have the opportunity, and retreat to the legion." Thus he fell fighting a few moments after, and saved his men by his own death.

LI. — Our soldiers, being hard pressed on every side, were dislodged from their position, with the loss of forty-six centurions; but the tenth legion, which had been posted in reserve on ground a little more level, checked the Gauls in their eager pursuit. It was supported by the cohorts of the thirteenth legion, which, being led from the smaller camp, had, under the command of Titus Sextius, occupied the higher ground. The legions, as soon as they reached the plain, halted and faced the enemy. Vercingetorix led back his men from the part of the hill within the fortifications. On that day little less than seven hundred of the soldiers were missing.

LII. — On the next day, Caesar, having called a meeting, censured the rashness and avarice of his soldiers, "In that they had judged for themselves how far they ought to proceed, or what they ought to do, and could not be kept back by the tribunes of the soldiers and the lieutenants;" and stated, "what the disadvantage of the ground could effect, what opinion he himself had entertained at Avaricum, when having surprised the enemy without either general or cavalry, he had given up a certain victory, lest even a trifling loss should occur in the contest owing to the disadvantage of position. That as much as he admired the greatness of their courage, since neither the fortifications of the camp, nor the height of the mountain, nor the wall of the town could retard them; in the same degree he censured their licentiousness and arrogance, because they thought that they knew more than their general concerning victory, and the issue of actions: and that he required in his soldiers forbearance and self-command, not less than valour and magnanimity."

LIII. — Having held this assembly, and having encouraged the soldiers at the conclusion of his speech, "That they should not be dispirited on this account, nor attribute to the valour of the enemy what the disadvantage of position had caused;" entertaining the same views of his departure that he had previously had, he led forth the legions from the camp, and drew up his army in order of battle in a suitable place. When Vercingetorix, nevertheless, would not descend to the level ground, a slight cavalry action, and that a successful one, having taken place, he led back his army into the camp. When he had done this, the next day, thinking that he had done enough to lower the pride of the Gauls, and to encourage the minds of his soldiers, he moved his camp in the direction of the Aedui. The enemy not even then pursuing us, on the third day he repaired the bridge over the river Allier, and led over his whole army.

LIV. — Having then held an interview with Viridomarus and Eporedorix the Aeduans, he learns that Litavicus had set out with all the cavalry to raise the Aedui; that it was necessary that they too should go before him to confirm the state in their allegiance. Although he now saw distinctly the treachery of the Aedui in many things, and was of opinion that the revolt of the entire state would be hastened by their departure; yet he thought that they should not be detained, lest he should appear either to offer an insult, or betray some suspicion of fear. He briefly states to them when departing his services towards the Aedui: in what a state and how humbled he had found them, driven into their towns, deprived of their lands, stripped of all their forces, a tribute imposed on them, and hostages wrested from them with the utmost insult; and to what condition and to what greatness he had raised them, [so much so] that they had not only recovered their former position, but seemed to surpass the dignity and influence of all the previous eras of their history. After giving these admonitions he dismissed them.

LV. — Noviodunum was a town of the Aedui, advantageously situated on the banks of the Loire. Caesar had conveyed hither all the hostages of Gaul, the corn, public money, a great part of his own baggage and that of his army; he had sent hither a great number of horses, which he had purchased in Italy and Spain on account of this war. When Eporedorix and Viridomarus came to this place, and received information of the disposition of the state, that Litavicus had been admitted by the Aedui into Bibracte, which is a town of the greatest importance among them, that Convictolitanis the chief magistrate and a great part of the senate had gone to meet him, that ambassadors had been publicly sent to Vercingetorix to negotiate a peace and alliance; they thought that so great an opportunity ought not to be neglected. Therefore, having put to the sword the garrison of Noviodunum and those who had assembled there for the purpose of trading or were on their march, they divided the money and horses among themselves; they took care that the hostages of the [different] states should be brought to Bibracte, to the chief magistrate; they burnt the town to prevent its being of any service to the Romans, as they were of opinion that they could not hold it; they carried away in their vessels whatever corn they could in the hurry; they destroyed the remainder, by [throwing it] into the river or setting it on fire; they themselves began to collect forces from the neighbouring country, to place guards and garrisons in different positions along the banks of the Loire, and to display the cavalry on all sides to strike terror into the Romans, [to try] if they could cut them off from a supply of provisions. In which expectation they were much aided, from the circumstance that the Loire had swollen to such a degree from the melting of the snows, that it did not seem capable of being forded at all.

LVI. — Caesar on being informed of these movements was of opinion that he ought to make haste, even if he should run some risk in completing the bridges, in order that he might engage before greater forces of the enemy should be collected in that place. For no one even then considered it an absolutely necessary act, that changing his design he should direct his march into the Province, both because the infamy and disgrace of the thing, and the intervening mount Cevennes, and the difficulty of the roads prevented him; and especially because he had serious apprehensions for the safety of Labienus whom he had detached, and those legions whom he had sent with him. Therefore, having made very long marches by day and night, he came to the river Loire, contrary to the expectation of all; and having by means of the cavalry found out a ford, suitable enough considering the emergency, of such depth that their arms and shoulders could be above water for supporting their accoutrements, he dispersed his cavalry in such a manner as to break the force of the current, and having confounded the enemy at the first sight, led his army across the river in safety; and finding corn and cattle in the fields, after refreshing his army with them, he determined to march into the country of the Senones.

LVII. — Whilst these things are being done by Caesar, Labienus, leaving at Agendicum the recruits who had lately arrived from Italy, to guard the baggage, marches with four legions to Lutetia (which is a town of the Parisii, situated on an island of the river Seine), whose arrival being discovered by the enemy, numerous forces arrived from the neighbouring states. The supreme command is entrusted to Camulogenus, one of the Aulerci, who, although almost worn out with age, was called to that honour on account of his extraordinary knowledge of military tactics. He, when he observed that there was a large marsh which communicated with the Seine, and rendered all that country impassable, encamped there, and determined to prevent our troops from passing it.

LVIII. — Labienus at first attempted to raise vineae, fill up the marsh with hurdles and clay, and secure a road. After he perceived that this was too difficult to accomplish, he issued in silence from his camp at the third watch, and reached Melodunum by the same route by which he came. This is a town of the Senones, situated on an island in the Seine, as we have just before observed of Lutetia. Having seized upon about fifty ships and quickly joined them together, and having placed soldiers in them, he intimidated by his unexpected arrival the inhabitants, of whom a great number had been called out to the war, and obtains possession of the town without a contest. Having repaired the bridge, which the enemy had broken down during the preceding days, he led over his army, and began to march along the banks of the river to Lutetia. The enemy, on learning the circumstance from those who had escaped from Melodunum, set fire to Lutetia, and order the bridges of that town to be broken down: they themselves set out from the marsh, and take their position on the banks of the Seine, over against Lutetia and opposite the camp of Labienus.

LIX. — Caesar was now reported to have departed from Gergovia; intelligence was likewise brought to them concerning the revolt of the Aedui, and a successful rising in Gaul; and that Caesar, having been prevented from prosecuting his journey and crossing the Loire, and having been compelled by the want of corn, had marched hastily to the province. But the Bellovaci, who had been previously disaffected of themselves, on learning the revolt of the Aedui, began to assemble forces and openly to prepare for war; Then Labienus, as the change in affairs was so great, thought that he must adopt a very different system from what he had previously intended, and he did not now think of making any new acquisitions, or of provoking the enemy to an action; but that he might bring back his army safe to Agendicum. For, on one side, the Bellovaci, a state which held the highest reputation for prowess in Gaul, were pressing on him; and Camulogenus, with a disciplined and well-equipped army, held the other side; moreover, a very great river separated and cut off the legions from the garrison and baggage. He saw that, in consequence of such great difficulties being thrown in his way, he must seek aid from his own energy of disposition.

LX. — Having, therefore, called a council of war a little before evening, he exhorted his soldiers to execute with diligence and energy such commands as he should give; he assigns the ships which he had brought from Melodunum to Roman knights, one to each, and orders them to fall down the river silently for four miles, at the end of the fourth watch, and there wait for him. He leaves the five cohorts, which he considered to be the most steady in action, to guard the camp; he orders the five remaining cohorts of the same legion to proceed a little after midnight up the river with all their baggage, in a great tumult. He collects also some small boats; and sends them in the same direction, with orders to make a loud noise in rowing. He himself, a little after, marched out in silence, and, at the head of three legions, seeks that place to which he had ordered the ships to be brought.

LXI. — When he had arrived there, the enemy's scouts, as they were stationed along every part of the river, not expecting an attack, because a great storm had suddenly arisen, were surprised by our soldiers: the infantry and cavalry are quickly transported, under the superintendence of the Roman knights, whom he had appointed to that office. Almost at the same time, a little before daylight, intelligence was given to the enemy that there was an unusual tumult in the camp of the Romans, and that a strong force was marching up the river, and that the sound of oars was distinctly heard in the same quarter, and that soldiers were being conveyed across in ships a little below. On hearing these things, because they were of opinion that the legions were passing in three different places, and that the entire army, being terrified by the revolt of the Aedui, were preparing for flight, they divided their forces also into three divisions. For leaving a guard opposite to the camp and sending a small body in the direction of Metiosedum, with orders to advance as far as the ships would proceed, they led the rest of their troops against Labienus.

LXII. — By day-break all our soldiers were brought across and the army of the enemy was in sight. Labienus, having encouraged his soldiers "to retain the memory of their ancient valour, and so many most successful actions, and imagine Caesar himself, under whose command they had so often routed the enemy, to be present," gives the signal for action. At the first onset the enemy are beaten and put to flight in the right wing, where the seventh legion stood: on the left wing, which position the twelfth legion held, although the first ranks fell transfixed by the javelins of the Romans, yet the rest resisted most bravely; nor did any one of them show the slightest intention of flying. Camulogenus, the general of the enemy, was present and encouraged his troops. But when the issue of the victory was still uncertain, and the circumstances which were taking place on the left wing were announced to the tribunes of the seventh legion, they faced about their legion to the enemy's rear and attacked it: not even then did any one retreat, but all were surrounded and slain. Camulogenus met the same fate. But those who were left as a guard opposite the camp of Labienus, when they heard that the battle was commenced, marched to aid their countrymen and take possession of a hill, but were unable to withstand the attack of the victorious soldiers. In this manner, mixed with their own fugitives, such as the woods and mountains did not shelter were cut to pieces by our cavalry. When this battle was finished, Labienus returns to Agendicum, where the baggage of the whole army had been left: from it he marched with all his forces to Caesar.

LXIII. — The revolt of the Aedui being known, the war grows more dangerous. Embassies are sent by them in all directions: as far as they can prevail by influence, authority, or money, they strive to excite the state [to revolt]. Having got possession of the hostages whom Caesar had deposited with them, they terrify the hesitating by putting them to death. The Aedui request Vercingetorix to come to them and communicate his plans of conducting the war. On obtaining this request they insist that the chief command should be assigned to them; and when the affair became a disputed question, a council of all Gaul is summoned to Bibracte. They come together in great numbers and from every quarter to the same place. The decision is left to the votes of the mass: all to a man approve of Vercingetorix as their general. The Remi, Lingones, and Treviri were absent from this meeting; the two former because they attached themselves to the alliance of Rome; the Treviri because they were very remote and were hard pressed by the Germans; which was also the reason of their being absent during the whole war, and their sending auxiliaries to neither party. The Aedui are highly indignant at being deprived of the chief command; they lament the change of fortune, and miss Caesar's indulgence towards them; however, after engaging in the war, they do not dare to pursue their own measures apart from the rest. Eporedorix and Viridomarus, youths of the greatest promise, submit reluctantly to Vercingetorix.

LXIV. — The latter demands hostages from the remaining states: nay, more, appointed a day for this proceeding; he orders all the cavalry, fifteen thousand in number, to quickly assemble here; he says that he will be content with the infantry which he had before, and would not tempt fortune nor come to a regular engagement; but since he had abundance of cavalry, it would be very easy for him to prevent the Romans from obtaining forage or corn, provided that they themselves should resolutely destroy their corn and set fire to their houses, by which sacrifice of private property they would evidently obtain perpetual dominion and freedom. After arranging these matters he levies ten thousand infantry on the Aedui and Segusiani, who border on our province: to these he adds eight hundred horse. He sets over them the brother of Eporedorix, and orders him to wage war against the Allobroges. On the other side he sends the Gabali and the nearest cantons of the Arverni against the Helvii; he likewise sends the Ruteni and Cadurci to lay waste the territories of the Volcae Arecomici. Besides, by secret messages and embassies, he tampers with the Allobroges, whose minds, he hopes, had not yet settled down after the excitement of the late war. To their nobles he promises money, and to their state the dominion of the whole province.

LXV. — The only guards provided against all these contingencies were twenty-two cohorts, which were collected from the entire province by Lucius Caesar, the lieutenant, and opposed to the enemy in every quarter. The Helvii, voluntarily engaging in battle with their neighbours, are defeated, and Caius Valerius Donotaurus, the son of Caburus, the principal man of the state, and several others, being slain, they are forced to retire within their towns and fortifications. The Allobroges, placing guards along the course of the Rhine, defend their frontiers with great vigilance and energy. Caesar, as he perceived that the enemy were superior in cavalry, and he himself could receive no aid from the province or Italy, while all communication was cut off, sends across the Rhine into Germany to those states which he had subdued in the preceding campaigns, and summons from them cavalry and the light-armed infantry, who were accustomed to engage among them. On their arrival, as they were mounted on unserviceable horses, he takes horses from the military tribunes and the rest, nay, even from the Roman knights and veterans, and distributes them among the Germans.

LXVI. — In the meantime, whilst these things are going on, the forces of the enemy from the Arverni, and the cavalry which had been demanded from all Gaul, meet together. A great number of these having been collected, when Caesar was marching into the country of the Sequani, through the confines of the Lingones, in order that he might the more easily render aid to the province, Vercingetorix encamped in three camps, about ten miles from the Romans: and having summoned the commanders of the cavalry to a council, he shows that the time of victory was come; that the Romans were fleeing into the province and leaving Gaul; that this was sufficient for obtaining immediate freedom; but was of little moment in acquiring peace and tranquillity for the future; for the Romans would return after assembling greater forces, and would not put an end to the war; therefore they should attack them on their march, when encumbered. If the infantry should [be obliged to] relieve their cavalry, and be retarded by doing so, the march could not be accomplished: if, abandoning their baggage, they should provide for their safety (a result which, he trusted, was more likely to ensue), they would lose both property and character. For as to the enemy's horse, they ought not to entertain a doubt that none of them would dare to advance beyond the main body. In order that they [the Gauls] may do so with greater spirit, he would marshal all their forces before the camp, and intimidate the enemy. The cavalry unanimously shout out, "That they ought to bind themselves by a most sacred oath, that he should not be received under a roof, nor have access to his children, parents, or wife, who shall not twice have ridden through the enemy's army."

LXVII. — This proposal receiving general approbation, and all being forced to take the oath, on the next day the cavalry were divided into three parts, and two of these divisions made a demonstration on our two flanks; while one in front began to obstruct our march. On this circumstance being announced, Caesar orders his cavalry also to form three divisions and charge the enemy. Then the action commences simultaneously in every part: the main body halts; the baggage is received within the ranks of the legions. If our men seemed to be distressed, or hard pressed in any quarter, Caesar usually ordered the troops to advance, and the army to wheel round in that quarter; which conduct retarded the enemy in the pursuit, and encouraged our men by the hope of support. At length the Germans, on the right wing, having gained the top of the hill, dislodge the enemy from their position and pursue them even as far as the river at which Vercingetorix with the infantry was stationed, and slay several of them. The rest, on observing this action, fearing lest they should be surrounded, betake themselves to flight. A slaughter ensues in every direction, and three of the noblest of the Audi are taken and brought to Caesar: Cotus, the commander of the cavalry, who had been engaged in the contest with Convictolitanis the last election, Cavarillus, who had held the command of the infantry after the revolt of Litavicus, and Eporedorix, under whose command the Aedui had engaged in war against the Sequani, before the arrival of Caesar.

LXVIII. — All his cavalry being routed, Vercingetorix led back his troops in the same order as he had arranged them before the camp, and immediately began to march to Alesia, which is a town of the Mandubii; and ordered the baggage to be speedily brought forth from the camp, and follow him closely. Caesar, having conveyed his baggage to the nearest hill, and having left two legions to guard it, pursued as far as the time of day would permit, and after slaying about three thousand of the rear of the enemy, encamped at Alesia on the next day. On reconnoitring the situation of the city, finding that the enemy were panic-stricken, because the cavalry in which they placed their chief reliance were beaten, he encouraged his men to endure the toil, and began to draw a line of circumvallation round Alesia.

LXIX. — The town itself was situated on the top of a hill, in a very lofty position, so that it did not appear likely to be taken, except by a regular siege. Two rivers, on two different sides, washed the foot of the hill. Before the town lay a plain of about three miles in length; on every other side hills at a moderate distance, and of an equal degree of height, surrounded the town. The army of the Gauls had filled all the space under the wall, comprising the part of the hill which looked to the rising sun, and had drawn in front a trench and a stone wall six feet high. The circuit of that fortification, which was commenced by the Romans, comprised eleven miles. The camp was pitched in a strong position, and twenty-three redoubts were raised in it, in which sentinels were placed by day, lest any sally should be made suddenly; and by night the same were occupied by watches and strong guards.

LXX. — The work having been begun, a cavalry action ensues in that plain, which we have already described as broken by hills, and extending three miles in length. The contest is maintained on both sides with the utmost vigour; Caesar sends the Germans to aid our troops when distressed, and draws up the legions in front of the camp, lest any sally should be suddenly made by the enemy's infantry. The courage of our men is increased by the additional support of the legions; the enemy being put to flight, hinder one another by their numbers, and as only the narrower gates were left open, are crowded together in them; then the Germans pursue them with vigour even to the fortifications. A great slaughter ensues; some leave their horses, and endeavour to cross the ditch and climb the wall. Caesar orders the legions which he had drawn up in front of the rampart to advance a little. The Gauls, who were within the fortifications, were no less panic-stricken, thinking that the enemy were coming that moment against them, and unanimously shout "to arms;" some in their alarm rush into the town; Vercingetorix orders the gates to be shut, lest the camp should be left undefended. The Germans retreat, after slaying many and taking several horses.

LXXI. — Vercingetorix adopts the design of sending away all his cavalry by night, before the fortifications should be completed by the Romans. He charges them when departing "that each of them should go to his respective state, and press for the war all who were old enough to bear arms; he states his own merits, and conjures them to consider his safety, and not surrender him, who had deserved so well of the general freedom, to the enemy for torture; he points out to them that, if they should be remiss, eighty thousand chosen men would perish with him; that, upon making a calculation, he had barely corn for thirty days, but could hold out a little longer by economy." After giving these instructions he silently dismisses the cavalry in the second watch, [on that side] where our works were not completed; he orders all the corn to be brought to himself; he ordains capital punishment to such as should not obey; he distributes among them, man by man, the cattle, great quantities of which had been driven there by the Mandubii; he began to measure out the corn sparingly, and by little and little; he receives into the town all the forces which he had posted in front of it. In this manner he prepares to await the succours from Gaul, and carry on the war.

LXXII. — Caesar, on learning these proceedings from the deserters and captives, adopted the following system of fortification; he dug a trench twenty feet deep, with perpendicular sides, in such a manner that the base of this trench should extend so far as the edges were apart at the top. He raised all his other works at a distance of four hundred feet from that ditch; [he did] that with this intention, lest (since he necessarily embraced so extensive an area, and the whole works could not be easily surrounded by a line of soldiers) a large number of the enemy should suddenly, or by night, sally against the fortifications; or lest they should by day cast weapons against our men while occupied with the works. Having left this interval, he drew two trenches fifteen feet broad, and of the same depth; the innermost of them, being in low and level ground, he filled with water conveyed from the river. Behind these he raised a rampart and wall twelve feet high: to this he added a parapet and battlements, with large stakes cut like stags' horns, projecting from the junction of the parapet and battlements, to prevent the enemy from scaling it, and surrounded the entire work with turrets, which were eighty feet distant from one another.

LXXIII. — It was necessary, at one and the same time, to procure timber [for the rampart], lay in supplies of corn, and raise also extensive fortifications, and the available troops were in consequence of this reduced in number, since they used to advance to some distance from the camp, and sometimes the Gauls endeavoured to attack our works, and to make a sally from the town by several gates and in great force. On which Caesar thought that further additions should be made to these works, in order that the fortifications might be defensible by a small number of soldiers. Having, therefore, cut down the trunks of trees or very thick branches, and having stripped their tops of the bark, and sharpened them into a point, he drew a continued trench everywhere five feet deep. These stakes being sunk into this trench, and fastened firmly at the bottom, to prevent the possibility of their being torn up, had their branches only projecting from the ground. There were five rows in connection with, and intersecting each other; and whoever entered within them were likely to impale themselves on very sharp stakes. The soldiers called these "*cippi.*" Before these, which were arranged in oblique rows in the form of a quincunx, pits three feet deep were dug, which gradually diminished in depth to the bottom. In these pits tapering stakes, of the thickness of a man's thigh, sharpened at the top and hardened in the fire, were sunk in such a manner as to project from the ground not more than four inches; at the same time for the purpose of giving them strength and stability, they were each filled with trampled clay to the height of one foot from the bottom: the rest of the pit was covered over with osiers and twigs, to conceal the deceit. Eight rows of this kind were dug, and were three feet distant from each other. They called this a lily from its resemblance to that flower. Stakes a foot long, with iron hooks attached to them, were entirely sunk in the ground before these, and were planted in every place at small intervals; these they called spurs.

LXXIV. — After completing these works, having selected as level ground as he could, considering the nature of the country, and having enclosed an area of fourteen miles, he constructed, against an external enemy, fortifications of the same kind in every respect, and separate from these, so that the guards of the fortifications could not be surrounded even by immense numbers, if such a circumstance should take place owing to the departure of the enemy's cavalry; and in order that the Roman soldiers might not be compelled to go out of the camp with great risk, he orders all to provide forage and corn for thirty days.

LXXV. — Whilst those things are carried on at Alesia, the Gauls, having convened a council of their chief nobility, determine that all who could bear arms should not be called out, which was the opinion of Vercingetorix, but that a fixed number should be levied from each state; lest, when so great a multitude assembled together, they could neither govern nor distinguish their men, nor have the means of supplying them with corn. They demand thirty-five thousand men from the Aedui and their dependents, the Segusiani, Ambivareti, and Aulerci Brannovices; an equal number from the Arverni in conjunction with the Eleuteti Cadurci, Gabali, and Velauni, who were accustomed to be under the command of the Arverni; twelve thousand each from the Senones, Sequani, Bituriges, Santones, Ruteni, and Carnutes; ten thousand from the Bellovaci; the same number from the Lemovici; eight thousand each from the Pictones, and Turoni, and Parisii, and Helvii; five thousand each from the Suessiones, Ambiani, Mediomatrici, Petrocorii, Nervii, Morini, and Nitiobriges; the same number from the Aulerci Cenomani; four thousand from the Atrebates; three thousand each from the Bellocassi, Lexovii, and Aulerci

Eburovices; thirty thousand from the Rauraci, and Boii; six thousand, from all the states together which border on the Atlantic, and which in their dialect are called Armoricae (in which number are comprehended the Curisolites, Rhedones, Ambibari, Caltes, Osismii, Lemovices, Veneti, and Unelli). Of these the Bellovaci did not contribute their number, as they said that they would wage war against the Romans on their own account, and at their own discretion, and would not obey the order of any one: however, at the request of Commius, they sent two thousand, in consideration of a tie of hospitality which subsisted between him and them.

LXXVI. — Caesar had, as we have previously narrated, availed himself of the faithful and valuable services of this Commius, in Britain, in former years: in consideration of which merits he had exempted from taxes his [Commius's] state, and had conferred on Commius himself the country of the Morini. Yet such was the unanimity of the Gauls in asserting their freedom, and recovering their ancient renown in war, that they were influenced neither by favours, nor by the recollection of private friendship; and all earnestly directed their energies and resources to that war, and collected eight thousand cavalry, and about two hundred and forty thousand infantry. These were reviewed in the country of the Aedui, and a calculation was made of their numbers: commanders were appointed: the supreme command is entrusted to Commius the Atrebatian, Viridomarus and Eporedorix the Aeduans, and Vergasillaunus the Arvernian, the cousin-german of Vercingetorix. To them are assigned men selected from each state, by whose advice the war should be conducted. All march to Alesia, sanguine and full of confidence: nor was there a single individual who imagined that the Romans could withstand the sight of such an immense host: especially in an action carried on both in front and rear, when [on the inside] the besieged would sally from the town and attack the enemy, and on the outside so great forces of cavalry and infantry would be seen.

LXXVII. — But those who were blockaded at Alesia, the day being past on which they had expected auxiliaries from their countrymen, and all their corn being consumed, ignorant of what was going on among the Aedui, convened an assembly and deliberated on the exigency of their situation. After various opinions had been expressed among them, some of which proposed a surrender, others a sally, whilst their strength would support it, the speech of Critognatus ought not to be omitted for its singular and detestable cruelty. He sprung from the noblest family among the Arverni, and possessing great influence, says, "I shall pay no attention to the opinion of those who call a most disgraceful surrender by the name of a capitulation; nor do I think that they ought to be considered as citizens, or summoned to the council. My business is with those who approve of a sally: in whose advice the memory of our ancient prowess seems to dwell in the opinion of you all. To be unable to bear privation for a short time is disgraceful cowardice, not true valour. Those who voluntarily offer themselves to death are more easily found than those who would calmly endure distress. And I would approve of this opinion (for honour is a powerful motive with me), could I foresee no other loss, save that of life: but let us, in adopting our design, look back on all Gaul, which we have stirred up to our aid. What courage do you think would our relatives and friends have, if eighty thousand men were butchered in one spot, supposing that they should be forced to come to an action almost over our corpses? Do not utterly deprive them of your aid, for they have spurned all thoughts of personal danger on account of your safety; nor by your folly, rashness, and cowardice, crush all Gaul and doom it to an eternal slavery. Do you doubt their fidelity and firmness because they have not come at the appointed day? What then? Do you suppose that the Romans are employed every day in the outer fortifications for mere amusement? If you cannot be assured by their despatches, since every avenue is blocked up, take the Romans as evidence that their approach is drawing near; since they, intimidated by alarm at this, labour night and day at their works. What, therefore, is my design? To do as our ancestors did in the war against the Cimbri and Teutones, which was by no means equally momentous; who, when driven into their towns, and oppressed by similar privations, supported life by the corpses of those who appeared useless for war on account of their age, and did not surrender to the enemy: and even if we had not a precedent for such cruel conduct, still I should consider it most glorious that one should be established, and delivered to posterity. For in what was that war like this? The Cimbri, after laying Gaul waste, and inflicting great calamities, at length departed from our country, and sought other lands; they left us our rights, laws, lands, and liberty. But what other motive or wish have the Romans, than, induced by envy, to settle in the lands and states of those whom they have learned by fame to be noble and powerful in war, and impose on them perpetual slavery? For they never have carried on wars on any other terms. But if you know not these things which are going on in distant countries, look to the neighbouring Gaul, which being reduced to the form of a province, stripped of its rights and laws, and subjected to Roman despotism, is oppressed by perpetual slavery."

LXXVIII. — When different opinions were expressed, they determined that those who, owing to age or ill health, were unserviceable for war, should depart from the town, and that themselves should try every expedient before they had recourse to the advice of Critognatus: however, that they would rather adopt that design, if circumstances should compel them and their allies should delay, than accept any terms of a surrender or peace. The Mandubii, who had admitted them into the town, are compelled to go forth with their wives and children. When these came to the Roman fortifications, weeping, they begged of the soldiers by every entreaty to receive them as slaves and relieve them with food. But Caesar, placing guards on the rampart, forbade them to be admitted.

LXXIX. — In the meantime, Commius and the rest of the leaders, to whom the supreme command had been intrusted, came with all their forces to Alesia, and having occupied the entire hill, encamp not more than a mile from our fortifications. The following day, having led forth their cavalry from the camp, they fill all that plain, which, we have related, extended three miles in length, and draw out their infantry a little from that place, and post them on the higher ground. The town Alesia commanded a view of the whole plain. The besieged run together when these auxiliaries were seen; mutual congratulations ensue, and the minds of all are elated with joy. Accordingly, drawing out their troops, they encamp before the town, and cover the nearest trench with hurdles and fill it up with earth, and make ready for a sally and every casualty.

LXXX. — Caesar, having stationed his army on both sides of the fortifications, in order that, if occasion should arise, each should hold and know his own post, orders the cavalry to issue forth from the camp and commence action. There was a commanding view from the entire camp, which occupied a ridge of hills; and the minds of all the soldiers anxiously awaited the issue of the battle. The Gauls had scattered archers and light-armed infantry here and there, among their cavalry, to give relief to their retreating troops, and sustain the impetuosity of our cavalry. Several of our soldiers were unexpectedly wounded by these, and left the battle. When the Gauls were confident that their countrymen were the conquerors in the action, and beheld our men hard pressed by numbers, both those who were hemmed in by the line of circumvallation and those who had come to aid them, supported the spirits of their men by shouts and yells from every quarter. As the action was carried on in sight of all, neither a brave nor cowardly act could be concealed; both the desire of praise and the fear of ignominy, urged on each party to valour. After fighting from noon almost to sunset, without victory inclining in favour of either, the Germans, on one side, made a charge against the enemy in a compact body, and drove them back; and, when they were put to flight, the archers were surrounded and cut to pieces. In other parts, likewise, our men pursued to the camp the retreating enemy, and did not give them an opportunity of rallying. But those who had come forth from Alesia returned into the town dejected and almost despairing of success.

LXXXI. — The Gauls, after the interval of a day, and after making, during that time, an immense number of hurdles, scaling ladders, and iron hooks, silently went forth from the camp at midnight and approached the fortifications in the plain. Raising a shout suddenly, that by this intimation those who were besieged in the town might learn their arrival, they began to cast down hurdles and dislodge our men from the rampart by slings, arrows, and stones, and executed the other movements which are requisite in storming. At the same time, Vercingetorix having heard the shout, gives the signal to his troops by a trumpet, and leads them forth from the town. Our troops, as each man's post had been assigned him some days before, man the fortifications; they intimidate the Gauls by slings, large stones, stakes which they had placed along the works, and bullets. All view being prevented by the darkness, many wounds are received on both sides; several missiles are thrown from the engines. But Marcus Antonius, and Caius Trebonius, the lieutenants, to whom the defence of these parts had been allotted, draughted troops from the redoubts which were more remote, and sent them to aid our troops, in whatever direction they understood that they were hard pressed.

LXXXII.—Whilst the Gauls were at a distance from the fortification, they did more execution, owing to the immense number of their weapons: after they came nearer, they either unawares empaled themselves on the spurs, or were pierced by the mural darts from the ramparts and towers, and thus perished. After receiving many wounds on all sides, and having forced no part of the works, when day drew nigh, fearing lest they should be surrounded by a sally made from the higher camp on the exposed flank, they retreated to their countrymen. But those within, whilst they bring forward those things which had been prepared by Vercingetorix for a sally, fill up the nearest trenches; having delayed a long time in executing these movements, they learned the retreat of their countrymen before they drew nigh to the fortifications. Thus they returned to the town without accomplishing their object.

LXXXIII. — The Gauls, having been twice repulsed with great loss, consult what they should do: they avail themselves of the information of those who were well acquainted with the country; from them they ascertain the position and fortification of the upper camp. There was, on the north side, a hill, which our men could not include in their works, on account of the extent of the circuit, and had necessarily made their camp in ground almost disadvantageous, and pretty steep. Caius Antistius Reginus, and Caius Caninius Rebilus, two of the lieutenants, with two legions, were in possession of this camp. The leaders of the enemy, having reconnoitred the country by their scouts, select from the entire army sixty thousand men; belonging to those states which bear the highest character for courage: they privately arrange among themselves what they wished to be done, and in what manner; they decide that the attack should take place when it should seem to be noon. They appoint over their forces Vergasillaunus, the Arvernian, one of the four generals, and a near relative of Vercingetorix. He, having issued from the camp at the first watch, and having almost completed his march a little before the dawn, hid himself behind the mountain, and ordered his soldiers to refresh themselves after their labour during the night. When noon now seemed to draw nigh, he marched hastily against that camp which we have mentioned before; and, at the same time, the cavalry began to approach the fortifications in the plain, and the rest of the forces to make a demonstration in front of the camp.

LXXXIV. — Vercingetorix, having beheld his countrymen from the citadel of Alesia, issues forth from the town; he brings forth from the camp long hooks, movable pent-houses, mural hooks, and other things, which he had prepared for the purpose of making a sally. They engage on all sides at once, and every expedient is adopted. They flocked to whatever part of the works seemed weakest. The army of the Romans is distributed along their extensive lines, and with difficulty meets the enemy in every quarter. The shouts which were raised by the combatants in their rear, had a great tendency to intimidate our men, because they perceived that their danger rested on the valour of others: for generally all evils which are distant most powerfully alarm men's minds.

LXXXV. — Caesar, having selected a commanding situation, sees distinctly whatever is going on in every quarter, and sends assistance to his troops when hard pressed. The idea uppermost in the minds of both parties is, that the present is the time in which they would have the fairest opportunity of making a struggle; the Gauls despairing of all safety, unless they should succeed in forcing the lines: the Romans expecting an end to all their labours if they should gain the day. The principal struggle is at the upper lines, to which, we have said, Vergasillaunus was sent. The least elevation of ground, added to a declivity, exercises a momentous influence. Some are casting missiles, others, forming a testudo, advance to the attack; fresh men by turns relieve the wearied. The earth, heaped up by all against the fortifications, gives the means of ascent to the Gauls, and covers those works which the Romans had concealed in the ground. Our men have no longer arms or strength.

LXXXVI. — Caesar, on observing these movements, sends Labienus with six cohorts to relieve his distressed soldiers: he orders him, if he should be unable to withstand them, to draw off the cohorts and make a sally; but not to do this except through necessity. He himself goes to the rest, and exhorts them not to succumb to the toil; he shows them that the fruits of all former engagements depend on that day and hour. The Gauls within, despairing of forcing the fortifications in the plains on account of the greatness of the works, attempt the places precipitous in ascent: hither they bring the engines which they had prepared; by the immense number of their missiles they dislodge the defenders from the turrets: they fill the ditches with clay and hurdles, then clear the way; they tear down the rampart and breast-work with hooks.

LXXXVII. — Caesar sends at first young Brutus, with six cohorts, and afterwards Caius Fabius, his lieutenant, with seven others: finally, as they fought more obstinately, he leads up fresh men to the assistance of his soldiers. After renewing the action, and repulsing the enemy, he marches in the direction in which he had sent Labienus, drafts four cohorts from the nearest redoubt, and orders part of the cavalry to follow him, and part to make the circuit of the external fortifications and attack the enemy in the rear. Labienus, when neither the ramparts or ditches could check the onset of the enemy, informs Caesar by messengers of what he intended to do. Caesar hastens to share in the action.

LXXXVIII. — His arrival being known from the colour of his robe, and the troops of cavalry, and the cohorts which he had ordered to follow him being seen, as these low and sloping grounds were plainly visible from the eminences, the enemy join battle. A shout being raised by both sides, it was succeeded by a general shout along the ramparts and whole line of fortifications. Our troops, laying aside their javelins, carry on the engagement with their swords. The cavalry is suddenly seen in the rear of the Gauls: the other cohorts advance rapidly; the enemy turn their backs; the cavalry intercept them in their flight, and a great slaughter ensues. Sedulius the general and chief of the Lemovices is slain; Vergasillaunus, the Arvernian, is taken alive in the flight, seventy-four military standards are brought to Caesar, and few out of so great a number return safe to their camp. The besieged, beholding from the town the slaughter and flight of their countrymen, despairing of safety, lead back their troops from the fortifications. A flight of the Gauls from their camp immediately ensues on hearing of this disaster, and had not the soldiers been wearied by sending frequent reinforcements, and the labour of the entire day, all the enemy's forces could have been destroyed. Immediately after midnight, the cavalry are sent out and overtake the rear, a great number are taken or cut to pieces, the rest by flight escape in different directions to their respective states. Vercingetorix, having convened a council the following day, declares, "That he had undertaken that war, not on account of his own exigencies, but on account of the general freedom; and since he must yield to fortune, he offered himself to them for either purpose, whether they should wish to atone to the Romans by his death, or surrender him alive." Ambassadors are sent to Caesar on this subject. He orders their arms to be surrendered, and their chieftains delivered up. He seated himself at the head of the lines in front of the camp, the Gallic chieftains are brought before him. They surrender Vercingetorix, and lay down their arms. Reserving the Aedui and Arverni, [to try] if he could gain over, through their influence, their respective states, he distributes one of the remaining captives to each soldier, throughout the entire army, as plunder.

XC. — After making these arrangements, he marches into the

[country of the] Aedui, and recovers that state. To this place ambassadors are sent by the Arverni, who promise that they will execute his commands. He demands a great number of hostages. He sends the legions to winter-quarters; he restores about twenty thousand captives to the Aedui and Arverni; he orders Titus Labienus to march into the [country of the] Sequani with two legions and the cavalry, and to him he attaches Marcus Sempronius Rutilus; he places Caius Fabius, and Lucius Minucius Basilus, with two legions in the country of the Remi, lest they should sustain any loss from the Bellovaci in their neighbourhood. He sends Caius Antistius Reginus into the [country of the] Ambivareti, Titus Sextius into the territories of the Bituriges, and Caius Caninius Rebilus into those of the Ruteni, with one legion each. He stations Quintus Tullius Cicero, and Publius Sulpicius among the Aedui at Cabillo and Matisco on the Saone, to procure supplies of corn. He himself determines to winter at Bibracte. A supplication of twenty days is decreed by the senate at Rome, on learning these successes from Caesar's despatches.

Book VIII

Continuation ascribed to Aulus Hirtius

Preface — Prevailed on by your continued solicitations, Balbus, I have engaged in a most difficult task, as my daily refusals appear to plead not my inability, but indolence, as an excuse. I have compiled a continuation of the Commentaries of our Caesar's Wars in Gaul, not indeed to be compared to his writings, which either precede or follow them; and recently, I have completed what he left imperfect after the transactions in Alexandria, to the end, not indeed of the civil broils, to which we see no issue, but of Caesar's life. I wish that those who may read them could know how unwillingly I undertook to write them, as then I might the more readily escape the imputation of folly and arrogance, in presuming to intrude among Caesar's writings. For it is agreed on all hands, that no composition was ever executed with so great care, that it is not exceeded in elegance by these Commentaries, which were published for the use of historians, that they might not want memoirs of such achievements; and they stand so high in the esteem of all men, that historians seem rather deprived of than furnished with materials. At which we have more reason to be surprised than other men; for they can only appreciate the elegance and correctness with which he finished them, while we know with what ease and expedition. Caesar possessed not only an uncommon flow of language and elegance of style, but also a thorough knowledge of the method of conveying his ideas. But I had not even the good fortune to share in the Alexandrian or African war; and though these were partly communicated to me by Caesar himself, in conversation, yet we listen with a different degree of attention to those things which strike us with admiration by their novelty, and those which we design to attest to posterity. But, in truth, whilst I urge every apology, that I may not be compared to Caesar, I incur the charge of vanity, by thinking it possible that I can in the judgment of any one be put in competition with him. Farewell.

I. — Gaul being entirely reduced, when Caesar having waged war incessantly during the former summer, wished to recruit his soldiers after so much fatigue, by repose in winter-quarters, news was brought him that several states were simultaneously renewing their hostile intentions, and forming combinations. For which a probable reason was assigned: namely, that the Gauls were convinced that they were not able to resist the Romans with any force they could collect in one place; and hoped that if several states made war in different places at the same time, the Roman army would neither have aid, nor time, nor forces, to prosecute them all: nor ought any single state to decline any inconveniences that might befall them, provided that by such delay the rest should be enabled to assert their liberty.

II. — That this notion might not be confirmed among the Gauls, Caesar left Marcus Antonius, his quaestor, in charge of his quarters, and set out himself with a guard of horse, the day before the kalends of January, from the town Bibracte, to the thirteenth legion, which he had stationed in the country of the Bituriges, not far from the territories of the Aedui, and joined to it the eleventh legion which was next it. Leaving two cohorts to guard the baggage, he leads the rest of his army into the most plentiful part of the country of the Bituriges; who, possessing an extensive territory and several towns, were not to be deterred, by a single legion quartered among them, from making warlike preparation, and forming combinations.

III. — By Caesar's sudden arrival, it happened, as it necessarily must, to an unprovided and dispersed people, that they were surprised by our horse, whilst cultivating the fields without any apprehensions, before they had time to fly to their towns. For the usual sign of an enemy's invasion, which is generally intimated by the burning of their towns, was forbidden by Caesar's orders: lest if he advanced far, forage and corn should become scarce, or the enemy be warned by the fires to make their escape. Many thousands being taken, as many of the Bituriges as were able to escape the first coming of the Romans, fled to the neighbouring states, relying either on private friendship, or public alliance. In vain; for Caesar, by hasty marches, anticipated them in every place, nor did he allow any state leisure to consider the safety of others, in preference to their own. By this activity, he both retained his friends in their loyalty, and by fear, obliged the wavering to accept offers of peace. Such offers being made to the Bituriges, when they perceived that through Caesar's clemency, an avenue was open to his friendship, and that the neighbouring states had given hostages, without incurring any punishment, and had been received under his protection, they did the same.

IV. — Caesar promises his soldiers, as a reward for their labour and patience, in cheerfully submitting to hardships from the severity of the winter, the difficulty of the roads, and the intolerable cold, two hundred sestertii each, and to every centurion two thousand, to be given instead of plunder; and sending his legions back to quarters, he himself returned on the fortieth day to Bibracte. Whilst he was dispensing justice there, the Bituriges send ambassadors to him, to entreat his aid against the Carnutes, who they complained had made war against them. Upon this intelligence, though he had not remained more than eighteen days in winter-quarters, he draws the fourteenth and sixth legion out of quarters on the Saone, where he had posted them as mentioned in a former Commentary to procure supplies of corn. With these two legions he marches in pursuit of the Carnutes.

V. — When the news of the approach of our army reached the enemy, the Carnutes, terrified by the sufferings of other states, deserted their villages and towns (which were small buildings, raised in a hurry, to meet the immediate necessity, in which they lived to shelter themselves against the winter, for, being lately conquered, they had lost several towns), and dispersed and fled. Caesar, unwilling to expose his soldiers to the violent storms that break out, especially at that season, took up his quarters at Genabum, a town of the Carnutes; and lodged his men in houses, partly belonging to the Gauls, and partly built to shelter the tents, and hastily covered with thatch. But the horse and auxiliaries he sends to all parts to which he was told the enemy had marched; and not without effect, as our men generally returned loaded with booty. The Carnutes, overpowered by the severity of the winter, and the fear of danger, and not daring to continue long in any place, as they were driven from their houses, and not finding sufficient protection in the woods, from the violence of the storms, after losing a considerable number of their men, disperse, and take refuge among the neighbouring states.

VI. — Caesar, being contented, at so severe a season, to disperse the gathering foes, and prevent any new war from breaking out, and being convinced, as far as reason could foresee, that no war of consequence could be set on foot in the summer campaign, stationed Caius Trebonius, with the two legions which he had with him, in quarters at Genabum: and being informed by frequent embassies from the Remi, that the Bellovaci (who exceed all the Gauls and Belgae in military prowess), and the neighbouring states, headed by Correus, one of the Bellovaci, and Comius, the Atrebatian, were raising an army, and assembling at a general rendezvous, designing with their united forces to invade the territories of the Suessiones, who were put under the patronage of the Remi: and moreover, considering that not only his honour, but his interest was concerned, that such of his allies, as deserved well of the republic, should suffer no calamity; he again draws the eleventh legion out of quarters and writes besides to Caius Fabius, to march with his two legions to the country of the Suessiones; and he sends to Trebonius for one of his two legions. Thus, as far as the convenience of the quarters, and the management of the war admitted, he laid the burden of the expedition on the legions by turns, without any intermission to his own toils.

VII. — As soon as his troops were collected, he marched against the Bellovaci: and pitching his camp in their territories, detached troops of horse all round the country, to take prisoners, from whom he might learn the enemy's plan. The horse, having executed his orders, bring him back word that but few were found in the houses: and that even these had not stayed at home to cultivate their lands (for the emigration was general from all parts), but had been sent back to watch our motions. Upon Caesar's inquiring from them, where the main body of the Bellovaci were posted, and what was their design: they made answer, "that all the Bellovaci, fit for carrying arms, had assembled in one place, and along with them the Ambiani, Aulerci, Caletes, Velocasses, and Atrebates, and that they had chosen for their camp an elevated position, surrounded by a dangerous morass: that they had conveyed all their baggage into the most remote woods: that several noblemen were united in the management of the war; but that the people were most inclined to be governed by Correus, because they knew that he had the strongest aversion to the name of the Roman people: that a few days before Comius had left the camp to engage the Germans to their aid whose nation bordered on theirs, and whose numbers were countless: that the Bellovaci had come to a resolution, with the consent of all the generals and the earnest desire of the people, if Caesar should come with only three legions, as was reported, to give him battle, that they might not be obliged to encounter his whole army on a future occasion, when they should be in a more wretched and distressed condition; but if he brought a stronger force, they intended to remain in the position they had chosen, and by ambuscade to prevent the Romans from getting forage (which at that season was both scarce and much scattered), corn, and other necessaries."

VIII. — When Caesar was convinced of the truth of this account from the concurring testimony of several persons, and perceived that the plans which were proposed were full of prudence, and very unlike the rash resolves of a barbarous people, he considered it incumbent on him to use every exertion, in order that the enemy might despise his small force and come to an action. For he had three veteran legions of distinguished valour, the seventh, eighth, and ninth. The eleventh consisted of chosen youth of great hopes, who had served eight campaigns, but who, compared with the others, had not yet acquired any great reputation for experience and valour. Calling therefore a council, and laying before it the intelligence which he had received, he encouraged his soldiers. In order if possible to entice the enemy to an engagement by the appearance of only three legions, he ranged his army in the following manner: that the seventh, eighth, and ninth legions should march before all the baggage; that then the eleventh should bring up the rear of the whole train of baggage (which however was but small, as is usual on such expeditions), so that the enemy could not get a sight of a greater number than they themselves were willing to encounter. By this disposition he formed his army almost into a square, and brought them within sight of the enemy sooner than was anticipated.

IX. — When the Gauls, whose bold resolutions had been reported to Caesar, saw the legions advance with a regular motion, drawn up in battle array; either from the danger of an engagement, or our sudden approach, or with the design of watching our movements, they drew up their forces before the camp, and did not quit the rising ground. Though Caesar wished to bring them to battle, yet being surprised to see so vast a host of the enemy, he encamped opposite to them, with a valley between them, deep rather than extensive. He ordered his camp to be fortified with a rampart twelve feet high, with breast-works built on it proportioned to its height; and two trenches, each fifteen feet broad, with perpendicular sides to be sunk: likewise several turrets, three stories high, to be raised, with a communication to each other by galleries laid across and covered over; which should be guarded in front by small parapets of osiers; that the enemy might be repulsed by two rows of soldiers. The one of whom, being more secure from danger by their height, might throw their darts with more daring and to a greater distance; the other, which was nearer the enemy, being stationed on the rampart, would be protected by their galleries from darts falling on their heads. At the entrance he erected gates and turrets of a considerable height.

X. — Caesar had a double design in this fortification; for he both hoped that the strength of his works, and his [apparent] fears would raise confidence in the barbarians; and when there should be occasion to make a distant excursion to get forage or corn, he saw that his camp would be secured by the works with a very small force. In the meantime there were frequent skirmishes across the marsh, a few on both sides sallying out between the two camps. Sometimes, however, our Gallic or German auxiliaries crossed the marsh, and furiously pursued the enemy; or on the other hand the enemy passed it and beat back our men. Moreover there happened in the course of our daily foraging, what must of necessity happen, when corn is to be collected by a few scattered men out of private houses, that our foragers dispersing in an intricate country were surrounded by the enemy; by which, though we suffered but an inconsiderable loss of cattle and servants, yet it raised foolish hopes in the barbarians; but more especially, because Comius, who I said had gone to get aid from the Germans, returned with some cavalry, and though the Germans were only 500, yet the barbarians were elated by their arrival.

XI. — Caesar, observing that the enemy kept for several days within their camp, which was well secured by a morass and its natural situation, and that it could not be assaulted without a dangerous engagement, nor the place enclosed with lines without an addition to his army, wrote to Trebonius to send with all despatch for the thirteenth legion which was in winter-quarters among the Bituriges under Titus Sextius, one of his lieutenants; and then to come to him by forced marches with the three legions. He himself sent the cavalry of the Remi, and Lingones, and other states, from whom he had required a vast number, to guard his foraging parties, and to support them in case of any sudden attack of the enemy.

XII. — As this continued for several days, and their vigilance was relaxed by custom (an effect which is generally produced by time), the Bellovaci, having made themselves acquainted with the daily stations of our horse, lie in ambush with a select body of foot in a place covered with woods; to it they sent their horse the next day, who were first to decoy our men into the ambuscade, and then when they were surrounded, to attack them. It was the lot of the Remi to fall into this snare, to whom that day had been allotted to perform this duty; for, having suddenly got sight of the enemy's cavalry, and despising their weakness, in consequence of their superior numbers, they pursued them too eagerly, and were surrounded on every side by the foot. Being by this means thrown into disorder they returned with more precipitation than is usual in cavalry actions, with the loss of Vertiscus, the governor of their state, and the general of their horse, who, though scarcely able to sit on horseback through years, neither, in accordance with the custom of the Gauls, pleaded his age in excuse for not accepting the command, nor would he suffer them to fight without him. The spirits of the barbarians were puffed up and inflated at the success of this battle, in killing the prince and general of the Remi; and our men were taught by this loss, to examine the country, and post their guards with more caution, and to be more moderate in pursuing a retreating enemy.

XIII. — In the meantime daily skirmishes take place continually in view of both camps; these were fought at the ford and pass of the morass. In one of these contests the Germans, whom Caesar had brought over the Rhine, to fight intermixed with the horse, having resolutely crossed the marsh, and slain the few who made resistance, and boldly pursued the rest, so terrified them, that not only those who were attacked hand to hand, or wounded at a distance, but even those who were stationed at a greater distance to support them, fled disgracefully; and being often beaten from the rising grounds, did not stop till they had retired into their camp, or some, impelled by fear, had fled farther. Their danger drew their whole army into such confusion, that it was difficult to judge whether they were more insolent after a slight advantage, or more dejected by a trifling calamity.

XIV. — After spending several days in the same camp, the guards of the Bellovaci, learning that Caius Trebonius was advancing nearer with his legions, and fearing a siege like that of Alesia, send off by night all who were disabled by age or infirmity, or unarmed, and along with them their whole baggage. Whilst they are preparing their disorderly and confused troop for march (for the Gauls are always attended by a vast multitude of waggons, even when they have very light baggage), being overtaken by daylight, they drew their forces out before their camp, to prevent the Romans attempting a pursuit before the line of their baggage had advanced to a considerable distance. But Caesar did not think it prudent to attack them when standing on their defence, with such a steep hill in their favour, nor keep his legions at such a distance that they could quit their post without danger: but, perceiving that his camp was divided from the enemy's by a deep morass, so difficult to cross that he could not pursue with expedition, and that the hill beyond the morass, which extended almost to the enemy's camp, was separated from it only by a small valley, he laid a bridge over the morass and led his army across, and soon reached the plain on the top of the hill, which was fortified on either side by a steep ascent. Having there drawn up his army in order of battle, he marched to the furthest hill, from which he could, with his engines, shower darts upon the thickest of the enemy.

XV. — The Gauls, confiding in the natural strength of their position, though they would not decline an engagement if the Romans attempted to ascend the hill, yet dared not divide their forces into small parties, lest they should be thrown into disorder by being dispersed, and therefore remained in order of battle. Caesar, perceiving that they persisted in their resolution, kept twenty cohorts in battle array, and, measuring out ground there for a camp, ordered it to be fortified. Having completed his works, he drew up his legions before the rampart and stationed the cavalry in certain positions, with their horses bridled. When the Bellovaci saw the Romans prepared to pursue them, and that they could not wait the whole night, or continue longer in the same place without provisions, they formed the following plan to secure a retreat. They handed to one another the bundles of straw and sticks on which they sat (for it is the custom of the Gauls to sit when drawn up in order of battle, as has been asserted in former commentaries), of which they had great plenty in their camp, and piled them in the front of their line; and at the close of the day, on a certain signal, set them all on fire at one and the same time. The continued blaze soon screened all their forces from the sight of the Romans, which no sooner happened than the barbarians fled with the greatest precipitation.

XVI. — Though Caesar could not perceive the retreat of the enemy for the intervention of the fire, yet, suspecting that they had adopted that method to favour their escape, he made his legions advance, and sent a party of horse to pursue them; but, apprehensive of an ambuscade, and that the enemy might remain in the same place and endeavour to draw our men into a disadvantageous situation, he advances himself but slowly. The horse, being afraid to venture into the smoke and dense line of flame, and those who were bold enough to attempt it being scarcely able to see their horses' heads, gave the enemy free liberty to retreat, through fear of an ambuscade. Thus, by a flight, full at once of cowardice and address, they advanced without any loss about ten miles, and encamped in a very strong position. From which, laying numerous ambuscades, both of horse and foot, they did considerable damage to the Roman foragers.

XVII. — After this had happened several times, Caesar discovered, from a certain prisoner, that Correus, the general of the Bellovaci, had selected six thousand of his bravest foot and a thousand horse, with which he designed to lie in ambush in a place to which he suspected the Romans would send to look for forage, on account of the abundance of corn and grass. Upon receiving information of their design Caesar drew out more legions than he usually did, and sent forward his cavalry as usual, to protect the foragers. With these he intermixed a guard of light infantry, and himself advanced with the legions as fast as he could.

XVIII. — The Gauls, placed in ambush, had chosen for the seat of action a level piece of ground, not more than a mile in extent, enclosed on every side by a thick wood or a very deep river, as by a toil, and this they surrounded. Our men, apprised of the enemy's design, marched in good order to the ground, ready both in heart and hand to give battle, and willing to hazard any engagement when the legions were at their back. On their approach, as Correus supposed that he had got an opportunity of effecting his purpose, he at first shows himself with a small party and attacks the foremost troops. Our men resolutely stood the charge, and did not crowd together in one place, as commonly happens from surprise in engagements between the horse, whose numbers prove injurious to themselves.

XIX. — When by the judicious arrangement of our forces only a few of our men fought by turns, and did not suffer themselves to be surrounded, the rest of the enemy broke out from the woods whilst Correus was engaged. The battle was maintained in different parts with great vigour, and continued for a long time undecided, till at length a body of foot gradually advanced from the woods in order of battle and forced our horse to give ground: the light infantry, which were sent before the legions to the assistance of the cavalry, soon came up, and, mixing with the horse, fought with great courage. The battle was for some time doubtful, but, as usually happens, our men, who stood the enemy's first charge, became superior from this very circumstance that, though suddenly attacked from an ambuscade, they had sustained no loss. In the meantime the legions were approaching, and several messengers arrived with notice to our men and the enemy that the [Roman] general was near at hand, with his forces in battle array. Upon this intelligence, our men, confiding in the support of the cohorts, fought most resolutely, fearing, lest if they should be slow in their operations they should let the legions participate in the glory of the conquest. The enemy lose courage and attempt to escape by different ways. In vain; for they were themselves entangled in that labyrinth in which they thought to entrap the Romans. Being defeated and put to the rout, and having lost the greater part of their men, they fled in consternation whither-soever chance carried them; some sought the woods, others the river, but were vigorously pursued by our men and put to the sword. Yet, in the meantime, Correus, unconquered by calamity, could not be prevailed on to quit the field and take refuge in the woods, or accept our offers of quarter, but, fighting courageously and wounding several, provoked our men, elated with victory, to discharge their weapons against him.

XX. — After this transaction, Caesar, having come up immediately after the battle, and imagining that the enemy, upon receiving the news of so great a defeat, would be so depressed that they would abandon their camp, which was not above eight miles distant from the scene of action, though he saw his passage obstructed by the river, yet he marched his army over and advanced. But the Bellovaci and the other states, being informed of the loss they had sustained by a few wounded men who having escaped by the shelter of the woods, had returned to them after the defeat, and learning that everything had turned out unfavourable, that Correus was slain, and the horse and most valiant of their foot cut off, imagined that the Romans were marching against them, and calling a council in haste by sound of trumpet, unanimously cry out to send ambassadors and hostages to Caesar.

XXI. — This proposal having met with general approbation, Comius the Atrebatian fled to those Germans from whom he had borrowed auxiliaries for that war. The rest instantly send ambassadors to Caesar; and requested that he would be contented with that punishment of his enemy, which if he had possessed the power to inflict on them before the engagement, when they were yet uninjured, they were persuaded from his usual clemency and mercy, he never would have inflicted; that the power of the Bellovaci was crushed by the cavalry action; that many thousands of their choicest foot had fallen, that scarce a man had escaped to bring the fatal news. That, however, the Bellovaci had derived from the battle one advantage, of some importance, considering their loss; that Correus, the author of the rebellion, and agitator of the people, was slain: for that whilst he lived, the senate had never equal influence in the state with the giddy populace.

XXII. — Caesar reminded the ambassadors who made these supplications, that the Bellovaci had at the same season the year before, in conjunction with other states of Gaul, undertaken a war, and that they had persevered the most obstinately of all in their purpose, and were not brought to a proper way of thinking by the submission of the rest; that he knew and was aware that the guilt of a crime was easily transferred to the dead; but that no one person could have such influence, as to be able by the feeble support of the multitude to raise a war and carry it on without the consent of the nobles, in opposition to the senate, and in despite of every virtuous man; however he was satisfied with the punishment which they had drawn upon themselves.

XXIII. — The night following the ambassadors bring back his answer to their countrymen, and prepare the hostages. Ambassadors flock in from the other states, which were waiting for the issue of the [war with the] Bellovaci: they give hostages, and receive his orders; all except Comius, whose fears restrained him from entrusting his safety to any person's honour. For the year before, while Caesar was holding the assizes in Hither Gaul, Titus Labienus, having discovered that Comius was tampering with the states, and raising a conspiracy against Caesar, thought he might punish his infidelity without perfidy; but judging that he would not come to his camp at his invitation, and unwilling to put him on his guard by the attempt, he sent Caius Volusenus Quadratus, with orders to have him put to death under pretence of a conference. To effect his purpose, he sent with him some chosen centurions. When they came to the conference, and Volusenus, as had been agreed on, had taken hold of Comius by the hand, and one of the centurions, as if surprised at so uncommon an incident, attempted to kill him, he was prevented by the friends of Comius, but wounded him severely in the head by the first blow. Swords were drawn on both sides, not so much with a design to fight as to effect an escape, our men believing that Comius had received a mortal stroke; and the Gauls, from the treachery which they had seen, dreading that a deeper design lay concealed. Upon this transaction, it was said that Comius made a resolution never to come within sight of any Roman.

XXIV. — When Caesar, having completely conquered the most warlike nations, perceived that there was now no state which could make preparations for war to oppose him, but that some were removing and fleeing from their country to avoid present subjection, he resolved to detach his army into different parts of the country. He kept with himself Marcus Antonius the quaestor, with the eleventh legion; Caius Fabius was detached with twenty-five cohorts into the remotest part of Gaul, because it was rumoured that some states had risen in arms, and he did not think that Caius Caninius Rebilus, who had the charge of that country, was strong enough to protect it with two legions. He ordered Titus Labienus to attend himself, and sent the twelfth legion which had been under him in winter-quarters, to Hither Gaul, to protect the Roman colonies, and prevent any loss by the inroads of barbarians, similar to that which had happened the year before to the Tergestines, who were cut off by a sudden depredation and attack. He himself marched to depopulate the country of Ambiorix, whom he had terrified and forced to fly, but despaired of being able to reduce under his power; but he thought it most consistent with his honour to waste his country both of inhabitants, cattle, and buildings, so that from the abhorrence of his countrymen, if fortune suffered any to survive, he might be excluded from a return to his state for the calamities which he had brought on it.

XXV. — After he had sent either his legions or auxiliaries through every part of Ambiorix's dominions, and wasted the whole country by sword, fire, and rapine, and had killed or taken prodigious numbers, he sent Labienus with two legions against the Treviri, whose state, from its vicinity to Germany, being engaged in constant war, differed but little from the Germans, in civilization and savage barbarity; and never continued in its allegiance, except when awed by the presence of his army.

XXVI. — In the meantime Caius Caninius, a lieutenant, having received information by letters and messages from Duracius, who had always continued in friendship to the Roman people, though a part of his state had revolted, that a great multitude of the enemy were in arms in the country of the Pictones, marched to the town Limonum. When he was approaching it, he was informed by some prisoners, that Duracius was shut up by several thousand men, under the command of Dumnacus, general of the Andes, and that Limonum was besieged, but not daring to face the enemy with his weak legions, he encamped in a strong position: Dumnacus, having notice of Caninius's approach, turned his whole force against the legions, and prepared to assault the Roman camp. But after spending several days in the attempt, and losing a considerable number of men, without being able to make a breach in any part of the works, he returned again to the siege of Limonum.

XXVII. — At the same time, Caius Fabius, a lieutenant, brings back many states to their allegiance, and confirms their submission by taking hostages; he was then informed by letters from Caninius, of the proceedings among the Pictones. Upon which he set off to bring assistance to Duracius. But Dumnacus hearing of the approach of Fabius, and despairing of safety, if at the same time he should be forced to withstand the Roman army without, and observe, and be under apprehension from the town's people, made a precipitate retreat from that place with all his forces. Nor did he think that he should be sufficiently secure from danger, unless he led his army across the Loire, which was too deep a river to pass except by a bridge. Though Fabius had not yet come within sight of the enemy, nor joined Caninius; yet being informed of the nature of the country, by persons acquainted with it, he judged it most likely that the enemy would take that way, which he found they did take. He therefore marched to that bridge with his army, and ordered his cavalry to advance no further before the legions, than that they could return to the same camp at night, without fatiguing their horses. Our horse pursued according to orders, and fell upon Dumnacus's rear, and attacking them on their march, while fleeing, dismayed, and laden with baggage, they slew a great number, and took a rich booty. Having executed the affair so successfully, they retired to the camp.

XXVIII. — The night following, Fabius sent his horse before him, with orders to engage the enemy, and delay their march till he himself should come up. That his orders might be faithfully performed,

Quintus Atius Varus, general of the horse, a man of uncommon spirit and skill, encouraged his men, and pursuing the enemy, disposed some of his troops in convenient places, and with the rest gave battle to the enemy. The enemy's cavalry made a bold stand, the foot relieving each other, and making a general halt, to assist their horse against ours. The battle was warmly contested. For our men, despising the enemy whom they had conquered the day before, and knowing that the legions were following them, animated both by the disgrace of retreating, and a desire of concluding the battle expeditiously by their own courage, fought most valiantly against the foot: and the enemy, imagining that no more forces would come against them, as they had experienced the day before, thought they had got a favourable opportunity of destroying our whole cavalry.

XXIX. — After the conflict had continued for some time with great violence, Dumnacus drew out his army in such a manner, that the foot should by turns assist the horse. Then the legions, marching in close order, came suddenly in sight of the enemy. At this sight, the barbarian horse were so astonished, and the foot so terrified, that breaking through the line of baggage, they betook themselves to flight with a loud shout, and in great disorder. But our horse, who a little before had vigorously engaged them, whilst they made resistance, being elated with joy at their victory, raising a shout on every side, poured round them as they ran, and as long as their horses had strength to pursue, or their arms to give a blow, so long did they continue the slaughter of the enemy in that battle, and having killed above twelve thousand men in arms, or such as threw away their arms through fear, they took their whole train of baggage.

XXX. — After this defeat, when it was ascertained that Drapes, a Senonian (who in the beginning of the revolt of Gaul, had collected from all quarters men of desperate fortunes, invited the slaves to liberty, called in the exiles of the whole kingdom, given an asylum to robbers, and intercepted the Roman baggage and provisions), was marching to the province with five thousand men, being all he could collect after the defeat, and that Luterius a Cadurcian who, as it has been observed in a former commentary, had designed to make an attack on the Province in the first revolt of Gaul, had formed a junction with him, Caius Caninius went in pursuit of them with two legions, lest great disgrace might be incurred from the fears or injuries done to the Province by the depredations of a band of desperate men.

XXXI. — Caius Fabius set off with the rest of the army to the Carnutes and those other states, whose forces he was informed had served as auxiliaries in that battle, which he fought against Dumnacus. For he had no doubt that they would be more submissive after their recent sufferings, but if respite and time were given them, they might be easily excited by the earnest solicitations of the same Dumnacus. On this occasion Fabius was extremely fortunate and expeditious in recovering the states. For the Carnutes, who, though often harassed had never mentioned peace, submitted and gave hostages: and the other states, which lie in the remotest parts of Gaul, adjoining the ocean, and which are called Armoricae, influenced by the example of the Carnutes, as soon as Fabius arrived with his legions, without delay comply with his command. Dumnacus, expelled from his own territories, wandering and skulking about, was forced to seek refuge by himself in the most remote parts of Gaul.

XXXII. — But Drapes in conjunction with Literius, knowing that Caninius was at hand with the legions, and that they themselves could not without certain destruction enter the boundaries of the province, whilst an army was in pursuit of them, and being no longer at liberty to roam up and down and pillage, halt in the country of the Cadurci, as Luterius had once in his prosperity possessed a powerful influence over the inhabitants, who were his countrymen, and being always the author of new projects, had considerable authority among the barbarians; with his own and Drapes' troops he seized Uxellodunum, a town formerly in vassalage to him and strongly fortified by its natural situation; and prevailed on the inhabitants to join him.

XXXIII. — After Caninius had rapidly marched to this place, and perceived that all parts of the town were secured by very craggy rocks, which it would be difficult for men in arms to climb even if they met with no resistance; and, moreover, observing that the town's people were possessed of effects, to a considerable amount, and that if they attempted to convey them away in a clandestine manner, they could not escape our horse, nor even our legions; he divided his forces into three parts, and pitched three camps on very high ground, with the intention of drawing lines round the town by degrees, as his forces could bear the fatigue.

XXXIV. — When the townsmen perceived his design, being terrified by the recollection of the distress at Alesia, they began to dread similar consequences from a siege; and above all Luterius, who had experienced that fatal event, cautioned them to make provision of corn; they therefore resolve by general consent to leave part of their troops behind, and set out with their light troops to bring in corn. The scheme having met with approbation, the following night Drapes and Luterius, leaving two thousand men in the garrison, marched out of the town with the rest. After a few days' stay in the country of the Cadurci (some of whom were disposed to assist them with corn, and others were unable to prevent their taking it) they collected a great store. Sometimes also attacks were made on our little forts by sallies at night. For this reason Caninius deferred drawing his works round the whole town, lest he should be unable to protect them when completed, or by disposing his garrisons in several places, should make them too weak.

XXXV. — Drapes and Luterius, having laid in a large supply of corn, occupy a position at about ten miles distance from the town, intending from it to convey the corn into the town by degrees. They chose each his respective department. Drapes stayed behind in the camp with part of the army to protect it; Luterius conveys the train with provisions into the town. Accordingly, having disposed guards here and there along the road, about the tenth hour of the night, he set out by narrow paths through the woods, to fetch the corn into the town. But their noise being heard by the sentinels of our camp, and the scouts which we had sent out, having brought an account of what was going on, Caninius instantly with the ready-armed cohorts from the nearest turrets made an attack on the convoy at the break of day. They, alarmed at so unexpected an evil, fled by different ways to their guard: which as soon as our men perceived, they fell with great fury on the escort, and did not allow a single man to be taken alive. Luterius escaped thence with a few followers, but did not return to the camp.

XXXVI. — After this success, Caninius learnt from some prisoners, that a part of the forces was encamped with Drapes, not more than ten miles off; which being confirmed by several, supposing that after the defeat of one general, the rest would be terrified, and might be easily conquered, he thought it a most fortunate event that none of the enemy had fled back from the slaughter to the camp, to give Drapes notice of the calamity which had befallen him. And as he could see no danger in making the attempt, he sent forward all his cavalry and the German foot, men of great activity, to the enemy's camp. He divides one legion among the three camps, and takes the other without baggage along with him. When he had advanced near the enemy, he was informed by scouts, which he had sent before him, that the enemy's camp, as is the custom of barbarians, was pitched low, near the banks of a river, and that the higher grounds were unoccupied: but that the German horse had made a sudden attack on them, and had begun the battle. Upon this intelligence, he marched up with his legion, armed and in order of battle. Then, on a signal being suddenly given on every side, our men took possession of the higher grounds. Upon this, the German horse observing the Roman colours, fought with great vigour. Immediately all the cohorts attack them on every side; and having either killed or made prisoners of them all, gained great booty. In that battle, Drapes himself was taken prisoner.

XXXVII. — Caninius, having accomplished the business so successfully, without having scarcely a man wounded, returned to besiege the town; and, having destroyed the enemy without, for fear of whom he had been prevented from strengthening his redoubts, and surrounding the enemy with his lines, he orders the work to be completed on every side. The next day, Caius Fabius came to join him with his forces, and took upon him the siege of one side.

XXXVIII. — In the meantime, Caesar left Caius Antonius in the country of the Bellovaci, with fifteen cohorts, that the Belgae might have no opportunity of forming new plans in the future. He himself visits the other states, demands a great number of hostages, and by his encouraging language allays the apprehensions of all. When he came to the Carnutes, in whose state he has in a former commentary mentioned that the war first broke out; observing, that from a consciousness of their guilt, they seemed to be in the greatest terror: to relieve the state the sooner from its fear, he demanded that Guturvatus, the promoter of that treason, and the instigator of that rebellion, should be delivered up to punishment. And though the latter did not dare to trust his life even to his own countrymen, yet such diligent search was made by them all, that he was soon brought to our camp. Caesar was forced to punish him, by the clamours of the soldiers, contrary to his natural humanity, for they alleged that all the dangers and losses incurred in that war, ought to be imputed to Guturvatus. Accordingly, he was whipped to death, and his head cut off.

XXXIX. — Here Caesar was informed by numerous letters from Caninius of what had happened to Drapes and Luterius, and in what conduct the town's people persisted: and though he despised the smallness of their numbers, yet he thought their obstinacy deserving a severe punishment, lest Gaul in general should adopt an idea that she did not want strength but perseverance to oppose the Romans; and lest the other states, relying on the advantage of situation, should follow their example and assert their liberty; especially as he knew that all the Gauls understood that his command was to continue but one summer longer, and if they could hold out for that time, that they would have no further danger to apprehend. He therefore left Quintus Calenus, one of his lieutenants behind him, with two legions, and instructions to follow him by regular marches. He hastened as much as he could with all the cavalry to Caninius.

XL. — Having arrived at Uxellodunum, contrary to the general expectation, and perceiving that the town was surrounded by the works, and that the enemy had no possible means of retiring from the assault, and being likewise informed by the deserters that the townsmen had abundance of corn; he endeavoured to prevent their getting water. A river divided the valley below, which almost surrounded the steep craggy mountain on which Uxellodunum was built. The nature of the ground prevented his turning the current; for it ran so low down at the foot of the mountain, that no drains could be sunk deep enough to draw it off in any direction. But the descent to it was so difficult, that if we made opposition, the besieged could neither come to the river, nor retire up the precipice without hazard of their lives. Caesar, perceiving the difficulty, disposed archers and slingers, and in some places, opposite to the easiest descents, placed engines, and attempted to hinder the townsmen from getting water at the river, which obliged them afterwards to go all to one place to procure water.

XLI. — Close under the walls of the town, a copious spring gushed out on that part, which for the space of nearly three hundred feet, was not surrounded by the river. Whilst every other person wished that the besieged could be debarred from this spring, Caesar alone saw that it could be effected, though not without great danger. Opposite to it he began to advance the vineae towards the mountain, and to throw up a mound, with great labour and continual skirmishing. For the townsmen ran down from the high ground, and fought without any risk, and wounded several of our men, yet they obstinately pushed on and were not deterred from moving forward the vineae, and from surmounting by their assiduity the difficulties of situation. At the same time they work mines, and move the crates and vineae to the source of the fountain. This was the only work which they could do without danger or suspicion. A mound sixty feet high was raised; on it was erected a turret of ten stories, not with the intention that it should be on a level with the wall (for that could not be effected by any works), but to rise above the top of the spring. When our engines began to play from it upon the paths that led to the fountain, and the townsmen could not go for water without danger, not only the cattle designed for food and the working cattle, but a great number of men also died of thirst.

XLII. — Alarmed at this calamity, the townsmen fill barrels with tallow, pitch, and dried wood; these they set on fire, and roll down on our works. At the same time, they fight most furiously, to deter the Romans, by the engagement and danger, from extinguishing the flames. Instantly a great blaze arose in the works. For whatever they threw down the precipice, striking against the vine and agger, communicated the fire to whatever was in the way. Our soldiers on the other hand, though they were engaged in a perilous sort of encounter, and labouring under the disadvantages of position, yet supported all with very great presence of mind. For the action happened in an elevated situation, and in sight of our army; and a great shout was raised on both sides; therefore every man faced the weapons of the enemy and the flames in as conspicuous a manner as he could, that his valour might be the better known and attested.

XLIII. — Caesar, observing that several of his men were wounded, ordered the cohorts to ascend the mountain on all sides, and, under pretence of assailing the walls, to raise a shout: at which the besieged being frightened, and not knowing what was going on in other places, call off their armed troops from attacking our works, and dispose them on the walls. Thus our men, without hazarding a battle, gained time partly to extinguish the works which had caught fire, and partly to cut off the communication. As the townsmen still continued to make an obstinate resistance, and even, after losing the greatest part of their forces by drought, persevered in their resolution: At last the veins of the spring were cut across by our mines, and turned from their course. By this their constant spring was suddenly dried up, which reduced them to such despair that they imagined that it was not done by the art of man, but the will of the gods; forced, therefore, by necessity, they at length submitted.

XLIV. — Caesar, being convinced that his lenity was known to all men, and being under no fears of being thought to act severely from a natural cruelty, and perceiving that there would be no end to his troubles if several states should attempt to rebel in like manner and in different places, resolved to deter others by inflicting an exemplary punishment on these. Accordingly he cut off the hands of those who had borne arms against him. Their lives he spared, that the punishment of their rebellion might be the more conspicuous. Drapes, who I have said was taken by Caninius, either through indignation and grief arising from his captivity, or through fear of severer punishments, abstained from food for several days, and thus perished. At the same time, Luterius, who, I have related, had escaped from the battle, having fallen into the hands of Epasnactus, an Arvernian (for he frequently changed his quarters, and threw himself on the honour of several persons, as he saw that he dare not remain long in one place, and was conscious how great an enemy he deserved to have in Caesar), was by this Epasnactus, the Arvernian, a sincere friend of the Roman people, delivered without any hesitation, a prisoner to Caesar.

XLV. — In the meantime, Labienus engages in a successful cavalry action among the Treviri; and, having killed several of them and of the Germans, who never refused their aid to any person against the Romans, he got their chiefs alive into his power, and, amongst them, Surus, an Aeduan, who was highly renowned both for his valour and birth, and was the only Aeduan that had continued in arms till that time. Caesar, being informed of this, and perceiving that he had met with good success in all parts of Gaul, and reflecting that, in former campaigns, [Celtic] Gaul had been conquered and subdued; but that he had never gone in person to Aquitania, but had made a conquest of it, in some degree, by Marcus Crassus, set out for it with two legions, designing to spend the latter part of the summer there. This affair he executed with his usual despatch and good fortune. For all the states of Aquitania sent ambassadors to him and delivered hostages. These affairs being concluded, he marched with a guard of cavalry towards Narbo, and drew off his army into winter-quarters by his lieutenants. He posted four legions in the country of the Belgae, under Marcus Antonius, Caius Trebonius, Publius Vatinius, and Quintus Tullius, his lieutenants. Two he detached to the Aedui, knowing them to have a very powerful influence throughout all Gaul. Two he placed among the Turoni, near the confines of the Carnutes, to keep in awe the entire tract of country bordering on the ocean; the other two he placed in the territories of the Lemovices, at a small distance from the Arverni, that no part of Gaul might be without an army. Having spent a few days in the province, he quickly ran through all the business of the assizes, settled all public disputes, and distributed rewards to the most deserving; for he had a good opportunity of learning how every person was disposed towards the republic during the general revolt of Gaul, which he had withstood by the fidelity and assistance of the Province.

XLVII. — Having finished these affairs, he returned to his legions among the Belgae and wintered at Nemetocenna: there he got intelligence that Comius, the Atrebatian had had an engagement with his cavalry. For when Antonius had gone into winter-quarters, and the state of the Atrebates continued in their allegiance, Comius, who, after that wound which I before mentioned, was always ready to join his countrymen upon every commotion, that they might not want a person to advise and head them in the management of the war, when his state submitted to the Romans, supported himself and his adherents on plunder by means of his cavalry, infested the roads, and intercepted several convoys which were bringing provisions to the Roman quarters.

XLVIII. — Caius Volusenus Quadratus was appointed commander of the horse under Antonius, to winter with him: Antonius sent him in pursuit of the enemy's cavalry; now Volusenus added to that valour which was pre-eminent in him, a great aversion to Comius, on which account he executed the more willingly the orders which he received. Having, therefore, laid ambuscades, he had several encounters with his cavalry and came off successful. At last, when a violent contest ensued, and Volusenus, through eagerness to intercept Comius, had obstinately pursued him with a small party; and Comius had, by the rapidity of his flight, drawn Volusenus to a considerable distance from his troops, he, on a sudden, appealed to the honour of all about him for assistance not to suffer the wound, which he had perfidiously received, to go without vengeance; and, wheeling his horse about, rode unguardedly before the rest up to the commander. All his horse following his example, made a few of our men turn their backs and pursued them. Comius, clapping spurs to his horse, rode up to Volusenus, and, pointing his lance, pierced him in the thigh with great force. When their commander was wounded, our men no longer hesitated to make resistance, and, facing about, beat back the enemy. When this occurred, several of the enemy, repulsed by the great impetuosity of our men, were wounded, and some were trampled to death in striving to escape, and some were made prisoners. Their general escaped this misfortune by the swiftness of his horse. Our commander, being severely wounded, so much so that he appeared to run the risk of losing his life, was carried back to the camp. But Comius, having either gratified his resentment, or, because he had lost the greatest part of his followers, sent ambassadors to Antonius, and assured him that he would give hostages as a security that he would go wherever Antonius should prescribe, and would comply with his orders, and only entreated that this concession should be made to his fears, that he should not be obliged to go into the presence of any Roman. As Antonius judged that his request originated in a just apprehension, he indulged him in it and accepted his hostages.

* * * * *

Caesar, I know, has made a separate commentary of each year's transactions, which I have not thought it necessary for me to do, because the following year, in which Lucius Paulus and Caius Marcellus were consuls, produced no remarkable occurrences in Gaul. But that no person may be left in ignorance of the place where Caesar and his army were at that time, I have thought proper to write a few words in addition to this commentary.

* * * * *

XLIX. — Caesar, whilst in winter-quarters in the country of the Belgae, made it his only business to keep the states in amity with him, and to give none either hopes of, or pretext for, a revolt. For nothing was further from his wishes than to be under the necessity of engaging in another war at his departure; lest, when he was drawing his army out of the country, any war should be left unfinished, which the Gauls would cheerfully undertake, when there was no immediate danger. Therefore, by treating the states with respect, making rich presents to the leading men, imposing no new burdens, and making the terms of their subjection lighter, he easily kept Gaul (already exhausted by so many unsuccessful battles) in obedience.

L. — When the winter-quarters were broken up, he himself, contrary to his usual practice, proceeded to Italy, by the longest possible stages, in order to visit the free towns and colonies, that he might recommend to them the petition of Marcus Antonius, his treasurer, for the priesthood. For he exerted his interest both cheerfully in favour of a man strongly attached to him, whom he had sent home before him to attend the election, and zealously to oppose the faction and power of a few men, who, by rejecting Marcus Antonius, wished to undermine Caesar's influence when going out of office. Though Caesar heard on the road, before he reached Italy, that he was created augur, yet he thought himself in honour bound to visit the free town and colonies, to return them thanks for rendering such service to Antonius by their presence in such great numbers [at the election], and at the same time to recommend to them himself, and his honour in his suit for the consulate the ensuing year. For his adversaries arrogantly boasted that Lucius Lentulus and Caius Marcellus had been appointed consuls, who would strip Caesar of all honour and dignity: and that the consulate had been injuriously taken from Sergius Galba, though he had been much superior in votes and interest, because he was united to Caesar, both by friendship, and by serving as lieutenant under him.

LI. — Caesar, on his arrival, was received by the principal towns and colonies with incredible respect and affection; for this was the first time he came since the war against united Gaul. Nothing was omitted which could be thought of for the ornament of the gates, roads, and every place through which Caesar was to pass. All the people with their children went out to meet him. Sacrifices were offered up in every quarter. The market places and temples were laid out with entertainments, as if anticipating the joy of a most splendid triumph. So great was the magnificence of the richer and zeal of the poorer ranks of the people.

LII. — When Caesar had gone through all the states of Cisalpine Gaul, he returned with the greatest haste to the army at Nemetocenna; and having ordered all his legions to march from winter-quarters to the territories of the Treviri, he went thither and reviewed them. He made Titus Labienus governor of Cisalpine Gaul, that he might be the more inclined to support him in his suit for the consulate. He himself made such journeys, as he thought would conduce to the health of his men by change of air; and though he was frequently told that Labienus was solicited by his enemies, and was assured that a scheme was in agitation by the contrivance of a few, that the senate should interpose their authority to deprive him of a part of his army; yet he neither gave credit to any story concerning Labienus, nor could be prevailed upon to do anything in opposition to the authority of the senate; for he thought that his cause would be easily gained by the free voice of the senators. For Caius Curio, one of the tribunes of the people, having undertaken to defend Caesar's cause and dignity, had often proposed to the senate, "that if the dread of Caesar's arms rendered any apprehensive, as Pompey's authority and arms were no less formidable to the forum, both should resign their command, and disband their armies. That then the city would be free, and enjoy its due rights." And he not only proposed this, but of himself called upon the senate to divide on the question. But the consuls and Pompey's friends interposed to prevent it; and regulating matters as they desired, they broke up the meeting.

LIII. — This testimony of the unanimous voice of the senate was very great, and consistent with their former conduct; for the preceding year, when Marcellus attacked Caesar's dignity, he proposed to the senate, contrary to the law of Pompey and Crassus, to dispose of Caesar's province, before the expiration of his command, and when the votes were called for, and Marcellus, who endeavoured to advance his own dignity, by raising envy against Caesar, wanted a division, the full senate went over to the opposite side. The spirit of Caesar's foes was not broken by this, but it taught them, that they ought to strengthen their interest by enlarging their connections, so as to force the senate to comply with whatever they resolved on.

LIV. — After this a decree was passed by the senate, that one legion should be sent by Pompey, and another by Caesar, to the Parthian war. But these two legions were evidently drawn from Caesar alone. For the first legion which Pompey sent to Caesar, he gave Caesar, as if it belonged to himself, though it was levied in Caesar's province. Caesar, however, though no one could doubt the design of his enemies, sent the legion back to Cneius Pompey, and in compliance with the decree of the senate, ordered the fifteenth, belonging to himself, and which was quartered in Cisalpine Gaul, to be delivered up. In its room he sent the thirteenth into Italy, to protect the garrisons from which he had drafted the fifteenth. He disposed his army in winter-quarters, placed Caius Trebonius, with four legions among the Belgae, and detached Caius Fabius, with four more, to the Aedui; for he thought that Gaul would be most secure if the Belgae, a people of the greatest valour, and the Aedui, who possessed the most powerful influence, were kept in awe by his armies.

LV. — He himself set out for Italy; where he was informed on his arrival, that the two legions sent home by him, and which by the senate's decree, should have been sent to the Parthian war, had been delivered over to Pompey, by Caius Marcellus the consul, and were retained in Italy. Although from this transaction it was evident to every one that war was designed against Caesar, yet he resolved to submit to any thing, as long as there were hopes left of deciding the dispute in an equitable manner, rather than have recourse to arms.

The Civil War

Book I

I. — When Caesar's letter was delivered to the consuls, they were with great difficulty, and a hard struggle of the tribunes, prevailed on to suffer it to be read in the senate; but the tribunes could not prevail, that any question should be put to the senate on the subject of the letter. The consuls put the question on the regulation of the state. Lucius Lentulus the consul promises that he will not fail the senate and republic, "if they declared their sentiments boldly and resolutely, but if they turned their regard to Caesar, and courted his favour, as they did on former occasions, he would adopt a plan for himself, and not submit to the authority of the senate: that he too had a means of regaining Caesar's favour and friendship." Scipio spoke to the same purport, "that it was Pompey's intention not to abandon the republic, if the senate would support him; but if they should hesitate and act without energy, they would in vain implore his aid, if they should require it hereafter."

II. — This speech of Scipio's, as the senate was convened in the city, and Pompey was near at hand, seemed to have fallen from the lips of Pompey himself. Some delivered their sentiments with more moderation, as Marcellus first, who in the beginning of his speech, said, "that the question ought not to be put to the senate on this matter, till levies were made throughout all Italy, and armies raised under whose protection the senate might freely and safely pass such resolutions as they thought proper": as Marcus Calidius afterwards, who was of opinion, "that Pompey should set out for his province, that there might be no cause for arms: that Caesar was naturally apprehensive as two legions were forced from him, that Pompey was retaining those troops, and keeping them near the city to do him injury": as Marcus Rufus, who followed Calidius almost word for word. They were all harshly rebuked by Lentulus, who peremptorily refused to propose Calidius's motion. Marcellus, overawed by his reproofs, retracted his opinion. Thus most of the senate, intimidated by the expressions of the consul, by the fears of a present army, and the threats of Pompey's friends, unwillingly and reluctantly adopted Scipio's opinion, that Caesar should disband his army by a certain day, and should he not do so, he should be considered as acting against the state. Marcus Antonius, and Quintus Cassius, tribunes of the people, interposed. The question was immediately put on their interposition. Violent opinions were expressed: whoever spoke with the greatest acrimony and cruelty, was most highly commended by Caesar's enemies.

III. — The senate having broken up in the evening, all who belonged to that order were summoned by Pompey. He applauded the forward, and secured their votes for the next day; the more moderate he reproved and excited against Caesar. Many veterans, from all parts, who had served in Pompey's armies, were invited to his standard by the hopes of rewards and promotions. Several officers belonging to the two legions, which had been delivered up by Caesar, were sent for. The city and the Comitium were crowded with tribunes, centurions, and veterans. All the consuls' friends, all Pompey's connections, all those who bore any ancient enmity to Caesar, were forced into the senate house. By their concourse and declarations the timid were awed, the irresolute confirmed, and the greater part deprived of the power of speaking their sentiments with freedom. Lucius Piso, the censor, offered to go to Caesar: as did likewise Lucius Roscius, the praetor, to inform him of these affairs, and require only six days' time to finish the business. Opinions were expressed by some to the effect that commissioners should be sent to Caesar to acquaint him with the senate's pleasure.

IV.— All these proposals were rejected, and opposition made to them all, in the speeches of the consul, Scipio, and Cato. An old grudge against Caesar and chagrin at a defeat actuated Cato. Lentulus was wrought upon by the magnitude of his debts, and the hopes of having the government of an army and provinces, and by the presents which he expected from such princes as should receive the title of friends of the Roman people, and boasted amongst his friends, that he would be a second Sylla, to whom the supreme authority should return. Similar hopes of a province and armies, which he expected to share with Pompey on account of his connection with him, urged on Scipio; and moreover, [he was influenced by] the fear of being called to trial, and the adulation and an ostentatious display of himself and his friends in power, who at that time had great influence in the republic, and courts of judicature. Pompey himself, incited by Caesar's enemies, because he was unwilling that any person should bear an equal degree of dignity, had wholly alienated himself from Caesar's friendship, and procured a reconciliation with their common enemies; the greatest part of whom he had himself brought upon Caesar during his affinity with him. At the same time, chagrined at the disgrace which he had incurred by converting the two legions from their expedition through Asia and Syria, to [augment] his own power and authority, he was anxious to bring matters to a war.

V.— For these reasons everything was done in a hasty and disorderly manner, and neither was time given to Caesar's relations to inform him [of the state of affairs] nor liberty to the tribunes of the people to deprecate their own danger, nor even to retain the last privilege, which Sylla had left them, the interposing their authority; but on the seventh day they were obliged to think of their own safety, which the most turbulent tribunes of the people were not accustomed to attend to, nor to fear being called to an account for their actions, till the eighth month. Recourse is had to that extreme and final decree of the senate (which was never resorted to even by daring proposers except when the city was in danger of being set on fire, or when the public safety was despaired of). "That the consuls, praetors, tribunes of the people, and proconsuls in the city should take care that the state received no injury." These decrees are dated the eighth day before the ides of January; therefore, in the first five days, on which the senate could meet, from the day on which Lentulus entered into his consulate, the two days of election excepted, the severest and most virulent decrees were passed against Caesar's government, and against those most illustrious characters, the tribunes of the people. The latter immediately made their escape from the city, and withdrew to Caesar, who was then at Ravenna, awaiting an answer to his moderate demands; [to see] if matters could be brought to a peaceful termination by any equitable act on the part of the enemies.

VI. — During the succeeding days the senate is convened outside the city. Pompey repeated the same things which he had declared through Scipio. He applauded the courage and firmness of the senate, acquainted them with his force, and told them that he had ten legions ready; that he was moreover informed and assured that Caesar's soldiers were disaffected, and that he could not persuade them to defend or even follow him. Motions were made in the senate concerning other matters; that levies should be made through all Italy; that Faustus Sylla should be sent as propraetor into Mauritania; that money should be granted to Pompey from the public treasury. It was also put to the vote that king Juba should be [honoured with the title of] friend and ally. But Marcellus said that he would not allow this motion for the present. Philip, one of the tribunes, stopped [the appointment of] Sylla; the resolutions respecting the other matters passed. The provinces, two of which were consular, the remainder praetorian, were decreed to private persons; Scipio got Syria, Lucius Domitius Gaul: Philip and Marcellus were omitted, from a private motive, and their lots were not even admitted. To the other provinces praetors were sent, nor was time granted as in former years, to refer to the people on their appointment, nor to make them take the usual oath, and march out of the city in a public manner, robed in the military habit, after offering their vows; a circumstance which had never before happened. Both the consuls leave the city, and private men had lictors in the city and capital, contrary to all precedents of former times. Levies were made throughout Italy, arms demanded, and money exacted from the municipal towns, and violently taken from the temples. All distinctions between things human and divine are confounded.

VII. — These things being made known to Caesar, he harangued his soldiers; he reminded them of the wrongs done to him at all times by his enemies, and complained that Pompey had been alienated from him and led astray by them through envy and a malicious opposition to his glory, though he had always favoured and promoted Pompey's honour and dignity. He complained that an innovation had been introduced into the republic, that the intercession of the tribunes, which had been restored a few years before by Sylla, was branded as a crime, and suppressed by force of arms; that Sylla, who had stripped the tribunes of every other power, had, nevertheless, left the privilege of intercession unrestrained; that Pompey, who pretended to restore what they had lost, had taken away the privileges which they formerly had; that whenever the senate decreed, "that the magistrates should take care that the republic sustained no injury" (by which words and decree the Roman people were obliged to repair to arms), it was only when pernicious laws were proposed; when the tribunes attempted violent measures; when the people seceded, and possessed themselves of the temples and eminences of the city; (and these instances of former times, he showed them were expiated by the fate of Saturninus and the Gracchi): that nothing of this kind was attempted now, nor even thought of: that no law was promulgated, no intrigue with the people going forward, no secession made; he exhorted them to defend from the malice of his enemies, the reputation and honour of that general, under whose command they had for nine years most successfully supported the state; fought many successful battles, and subdued all Gaul and Germany." The soldiers of the thirteenth legion, which was present (for in the beginning of the disturbances he had called it out, his other legions not having yet arrived), all cry out that they are ready to defend their general, and the tribunes of the commons, from all injuries.

VIII. — Having made himself acquainted with the disposition of his soldiers, Caesar set off with that legion to Ariminum, and there met the tribunes, who had fled to him for protection; he called his other legions from winter-quarters, and ordered them to follow him. Thither came Lucius Caesar, a young man, whose father was a lieutenant general under Caesar. He, after concluding the rest of his speech, and stating for what purpose he had come, told Caesar that he had commands of a private nature for him from Pompey; that Pompey wished to clear himself to Caesar, lest he should impute those actions which he did for the republic, to a design of affronting him; that he had ever preferred the interest of the state to his own private connections; that Caesar, too, for his own honour, ought to sacrifice his desires and resentment to the public good, and not vent his anger so violently against his enemies, lest in his hopes of injuring them, he should injure the republic. He spoke a few words to the same purport from himself, in addition to Pompey's apology. Roscius, the praetor, conferred with Caesar almost in the same words, and on the same subject, and declared that Pompey had empowered him to do so.

IX. — Though these things seemed to have no tendency towards redressing his injuries, yet having got proper persons by whom he could communicate his wishes to Pompey; he required of them both, that as they had conveyed Pompey's demands to him, they should not refuse to convey his demands to Pompey; if by so little trouble they could terminate a great dispute, and liberate all Italy from her fears. "That the honour of the republic had ever been his first object, and dearer to him than life; that he was chagrined, that the favour of the Roman people was wrested from him by the injurious reports of his enemies; that he was deprived of a half-year's command, and dragged back to the city, though the people had ordered that regard should be paid to his suit for the consulate at the next election, though he was not present; that, however, he had patiently submitted to this loss of honour for the sake of the republic; that when he wrote letters to the senate, requiring that all persons should resign the command of their armies, he did not obtain even that request; that levies were made throughout Italy; that the two legions which had been taken from him, under the pretence of the Parthian war, were kept at home, and that the state was in arms. To what did all these things tend, unless to his ruin? But, nevertheless, he was ready to condescend to any terms, and to endure everything for the sake of the republic. Let Pompey go to his own province; let them both disband their armies; let all persons in Italy lay down their arms; let all fears be removed from the city; let free elections, and the whole republic be resigned to the direction of the senate and Roman people. That these things might be the more easily performed, and conditions secured and confirmed by oath, either let Pompey come to Caesar, or allow Caesar to go to him; it might be that all their disputes would be settled by an interview."

X. — Roscius and Lucius Caesar, having received this message, went to Capua, where they met the consuls and Pompey, and declared to them Caesar's terms. Having deliberated on the matter, they replied, and sent written proposals to him by the same persons, the purport of which was, that Caesar should return into Gaul, leave Ariminum, and disband his army: if he complied with this, that Pompey would go to Spain. In the meantime, until security was given that Caesar would perform his promises, that the consuls and Pompey would not give over their levies.

XI. — It was not an equitable proposal, to require that Caesar should quit Ariminum and return to his province; but that he [Pompey] should himself retain his province and the legions that belonged to another, and desire that Caesar's army should be disbanded, whilst he himself was making new levies: and that he should merely promise to go to his province, without naming the day on which he would set out; so that if he should not set out till after Caesar's consulate expired, yet he would not appear bound by any religious scruples about asserting a falsehood. But his not granting time for a conference, nor promising to set out to meet him, made the expectation of peace appear very hopeless. Caesar, therefore, sent Marcus Antonius, with five cohorts from Ariminum to Arretium; he himself stayed at Ariminum with two legions, with the intention of raising levies there. He secured Pisaurus, Fanum, and Ancona, with a cohort each.

XII. — In the meantime, being informed that Thermus the praetor was in possession of Iguvium, with five cohorts, and was fortifying the town, but that the affections of all the inhabitants were very well inclined towards himself; he detached Curio with three cohorts, which he had at Ariminum and Pisaurus. Upon notice of his approach, Thermus, distrusting the affections of the townsmen, drew his cohorts out of it, and made his escape; his soldiers deserted him on the road, and returned home. Curio recovered Iguvium, with the cheerful concurrence of all the inhabitants. Caesar, having received an account of this, and relying on the affections of the municipal towns, drafted all the cohorts of the thirteenth legion from the garrisons, and set out for Auximum, a town into which Attius had brought his cohorts, and of which he had taken possession, and from which he had sent senators round about the country of Picenum, to raise new levies.

XIII. — Upon news of Caesar's approach, the senate of Auximum went in a body to Attius Varus; and told him that it was not a subject for them to determine upon: yet neither they, nor the rest of the freemen would suffer Caius Caesar, a general, who had merited so well of the republic, after performing such great achievements, to be excluded from their town and walls; wherefore he ought to pay some regard to the opinion of posterity, and his own danger. Alarmed at this declaration, Attius Varus drew out of the town the garrison which he had introduced, and fled. A few of Caesar's front rank having pursued him, obliged him to halt, and when the battle began, Varus is deserted by his troops: some of them disperse to their homes, the rest come over to Caesar; and along with them, Lucius Pupius, the chief centurion, is taken prisoner and brought to Caesar. He had held the same rank before in Cneius Pompey's army. But Caesar applauded the soldiers of Attius, set Pupius at liberty, returned thanks to the people of Auximum, and promised to be grateful for their conduct.

XIV. — Intelligence of this being brought to Rome, so great a panic spread on a sudden that when Lentulus, the consul, came to open the treasury, to deliver money to Pompey by the senate's decree, immediately on opening the hallowed door he fled from the city. For it was falsely rumoured that Caesar was approaching, and that his cavalry were already at the gates. Marcellus, his colleague, followed him, and so did most of the magistrates. Cneius Pompey had left the city the day before, and was on his march to those legions which he had received from Caesar, and had disposed in winter-quarters in Apulia. The levies were stopped within the city. No place on this side of Capua was thought secure. At Capua they first began to take courage and to rally, and determined to raise levies in the colonies, which had been sent thither by the Julian law: and Lentulus brought into the public market-place the gladiators which Caesar maintained there for the entertainment of the people, and confirmed them in their liberty, and gave them horses and ordered them to attend him; but afterwards, being warned by his friends that this action was censured by the judgment of all, he distributed them among the slaves of the districts of Campania, to keep guard there.

XV. — Caesar, having moved forward from Auximum, traversed the whole country of Picenum. All the governors in these countries most cheerfully received him, and aided his army with every necessary. Ambassadors came to him even from Cingulum, a town which Labienus had laid out and built at his own expense, and offered most earnestly to comply with his orders. He demanded soldiers: they sent them. In the meantime, the twelfth legion came to join Caesar; with these two he marched to Asculum, the chief town of Picenum. Lentulus Spinther occupied that town with ten cohorts; but, on being informed of Caesar's approach, he fled from the town, and, in attempting to bring off his cohorts with him, was deserted by a great part of his men. Being left on the road with a small number, he fell in with Vibullius Rufus, who was sent by Pompey into Picenum to confirm the people [in their allegiance]. Vibullius, being informed by him of the transactions in Picenum, takes his soldiers from him and dismisses him. He collects, likewise, from the neighbouring countries, as many cohorts as he can from Pompey's new levies. Amongst them he meets with Ulcilles Hirrus fleeing from Camerinum, with six cohorts, which he had in the garrison there; by a junction with which he made up thirteen cohorts. With them he marched by hasty journeys to Corfinium, to Domitius Aenobarbus, and informed him that Caesar was advancing with two legions. Domitius had collected about twenty cohorts from Alba, and the Marsians, Pelignians, and neighbouring states.

XVI. — Caesar, having recovered Asculum and driven out Lentulus, ordered the soldiers that had deserted from him to be sought out and a muster to be made; and, having delayed for one day there to provide corn, he marched to Corfinium. On his approach, five cohorts, sent by Domitius from the town, were breaking down a bridge which was over the river, at three miles' distance from it. An engagement taking place there with Caesar's advanced-guard, Domitius's men were quickly beaten off from the bridge and retreated precipitately into the town. Caesar, having marched his legions over, halted before the town and encamped close by the walls.

XVII. — Domitius, upon observing this, sent messengers well acquainted with the country, encouraged by a promise of being amply rewarded, with despatches to Pompey to Apulia, to beg and entreat him to come to his assistance. That Caesar could be easily enclosed by the two armies, through the narrowness of the country, and prevented from obtaining supplies: unless he did so, that he and upwards of thirty cohorts, and a great number of senators and Roman knights, would be in extreme danger. In the meantime he encouraged his troops, disposed engines on the walls, and assigned to each man a particular part of the city to defend. In a speech to the soldiers he promised them lands out of his own estate; to every private soldier four acres, and a corresponding share to the centurions and veterans.

XVIII. — In the meantime, word was brought to Caesar that the people of Sulmo, a town about seven miles distant from Corfinium, were ready to obey his orders, but were prevented by Quintus Lucretius, a senator, and Attius, a Pelignian, who were in possession of the town with a garrison of seven cohorts. He sent Marcus Antonius thither, with five cohorts of the eighth legion. The inhabitants, as soon as they saw our standards, threw open their gates, and all the people, both citizens and soldiers, went out to meet and welcome Antonius. Lucretius and Attius leaped off the walls. Attius, being brought before Antonius, begged that he might be sent to Caesar. Antonius returned the same day on which he had set out with the cohorts and Attius. Caesar added these cohorts to his own army, and sent Attius away in safety. The three first days Caesar employed in fortifying his camp with strong works, in bringing in corn from the neighbouring free towns, and waiting for the rest of his forces. Within the three days the eighth legion came to him, and twenty-two cohorts of the new levies in Gaul, and about three hundred horse from the king of Noricum. On their arrival he made a second camp on another part of the town, and gave the command of it to Curio. He determined to surround the town with a rampart and turrets during the remainder of the time. Nearly at the time when the greatest part of the work was completed, all the messengers sent to Pompey returned.

XIX. — Having read Pompey's letter, Domitius, concealing the truth, gave out in council that Pompey would speedily come to their assistance; and encouraged them not to despond, but to provide everything necessary for the defence of the town. He held private conferences with a few of his most intimate friends, and determined on the design of fleeing. As Domitius's countenance did not agree with his words, and he did everything with more confusion and fear than he had shown on the preceding days, and as he had several private meetings with his friends, contrary to his usual practice, in order to take their advice, and as he avoided all public councils and assemblies of the people, the truth could be no longer hid nor dissembled; for Pompey had written back in answer, "That he would not put matters to the last hazard; that Domitius had retreated into the town of Corfinium, without either his advice or consent. Therefore, if any opportunity should offer, he [Domitius] should come to him with the whole force." But the blockade and works round the town prevented his escape.

XX. — Domitius's design being noised abroad, the soldiers in Corfinium early in the evening began to mutiny, and held a conference with each other by their tribunes and centurions, and the most respectable amongst themselves: "that they were besieged by Caesar; that his works and fortifications were almost finished; that their general, Domitius, on whose hopes and expectations they had confided, had thrown them off, and was meditating his own escape; that they ought to provide for their own safety." At first the Marsians differed in opinion, and possessed themselves of that part of the town which they thought the strongest. And so violent a dispute arose between them, that they attempted to fight and decide it by arms. However, in a little time, by messengers sent from one side to the other, they were informed of Domitius's meditated flight, of which they were previously ignorant. Therefore they all with one consent brought Domitius into public view, gathered round him, and guarded him; and sent deputies out of their number to Caesar, to say that they were ready to throw open their gates, to do whatever he should order, and to deliver up Domitius alive into his hands.

XXI. — Upon intelligence of these matters, though Caesar thought it of great consequence to become master of the town as soon as possible, and to transfer the cohorts to his own camp, lest any change should be wrought on their inclinations by bribes, encouragement, or fictitious messages, because in war great events are often brought about by trifling circumstances; yet, dreading lest the town should be plundered by the soldiers entering into it, and taking advantage of the darkness of the night, he commended the persons who came to him, and sent them back to the town, and ordered the gates and walls to be secured. He disposed his soldiers on the works, which he had begun, not at certain intervals, as was his practice before, but in one continued range of sentinels and stations, so that they touched each other, and formed a circle round the whole fortification; he ordered the tribunes and general officers to ride round; and exhorted them not only to be on their guard against sallies from the town, but also to watch that no single person should get out privately. Nor was any man so negligent or drowsy as to sleep that night. To so great height was their expectation raised, that they were carried away, heart and soul, each to different objects, what would become of the Corfinians, what of Domitius, what of Lentulus, what of the rest; what event would be the consequence of another.

XXII. — About the fourth watch, Lentulus Spinther said to our sentinels and guards from the walls, that he desired to have an interview with Caesar, if permission were given him. Having obtained it, he was escorted out of town; nor did the soldiers of Domitius leave him till they brought him into Caesar's presence. He pleaded with Caesar for his life, and entreated him to spare him, and reminded him of their former friendship; and acknowledged that Caesar's favours to him were very great; in that through his interest he had been admitted into the college of priests; in that after his praetorship he had been appointed to the government of Spain; in that he had been assisted by him in his suit for the consulate. Caesar interrupted him in his speech, and told him, "that he had not left his province to do mischief [to any man], but to protect himself from the injuries of his enemies; to restore to their dignity the tribunes of the people who had been driven out of the city on his account, and to assert his own liberty, and that of the Roman people, who were oppressed by a few factious men." Encouraged by this address, Lentulus begged leave to return to the town, that the security which he had obtained for himself might be an encouragement to the rest to hope for theirs; saying that some were so terrified that they were induced to make desperate attempts on their own lives. Leave being granted him, he departed.

XXIII. — When day appeared Caesar ordered all the senators and their children, the tribunes of the soldiers, and the Roman knights, to be brought before him. Among the persons of senatorial rank were Lucius Domitius, Publius Lentulus Spinther, Lucius Vibullius Rufus, Sextus Quintilius Varus, the quaestor, and Lucius Rubrius, besides the son of Domitius, and several other young men, and a great number of Roman knights and burgesses, whom Domitius had summoned from the municipal towns. When they were brought before him he protected them from the insolence and taunts of the soldiers; told them in few words that they had not made him a grateful return, on their part, for his very extraordinary kindness to them, and dismissed them all in safety. Sixty sestertia, which Domitius had brought with him and lodged in the public treasury, being brought to Caesar by the magistrates of Corfinium, he gave them back to Domitius, that he might not appear more moderate with respect to the life of men than in money matters, though he knew that it was public money, and had been given by Pompey to pay his army. He ordered Domitius's soldiers to take the oath to himself, and that day decamped and performed the regular march. He stayed only seven days before Corfinium, and marched into Apulia through the country of the Marrucinians, Frentanians, and Larinates.

XXIV. — Pompey, being informed of what had passed at Corfinium, marches from Luceria to Canusium, and thence to Brundusium. He orders all the forces raised everywhere by the new levies to repair to him.

He gives arms to the slaves that attended the flocks, and appoints horses for them. Of these he made up about three hundred horse. Lucius, the praetor, fled from Alba, with six cohorts: Rutilus Lupus, the praetor, from Tarracina, with three. These having descried Caesar's cavalry at a distance, which were commanded by Bivius Curius, and having deserted the praetor, carried their colours to Curius and went over to him. In like manner during the rest of his march, several cohorts fell in with the main body of Caesar's army, others with his horse. Cneius Magius, from Cremona, engineer-general to Pompey, was taken prisoner on the road and brought to Caesar, but sent back by him to Pompey with this message: "As hitherto he had not been allowed an interview, and was now on his march to him at Brundusium, that it deeply concerned the commonwealth and general safety that he should have an interview with Pompey; and that the same advantage could not be gained at a great distance when the proposals were conveyed to them by others, as if terms were argued by them both in person."

XXV. — Having delivered this message he marched to Brundusium with six legions, four of them veterans: the rest those which he had raised in the late levy and completed on his march, for he had sent all Domitius's cohorts immediately from Corfinium to Sicily. He discovered that the consuls were gone to Dyrrachium with a considerable part of the army, and that Pompey remained at Brundusium with twenty cohorts; but could not find out, for a certainty, whether Pompey stayed behind to keep possession of Brundusium, that he might the more easily command the whole Adriatic sea, with the extremities of Italy and the coast of Greece, and be able to conduct the war on either side of it, or whether he remained there for want of shipping; and, being afraid that Pompey would come to the conclusion that he ought not to relinquish Italy, he determined to deprive him of the means of communication afforded by the harbour of Brundusium. The plan of his work was as follows: — Where the mouth of the port was narrowest he threw up a mole of earth on either side, because in these places the sea was shallow. Having gone out so far that the mole could not be continued in the deep water, he fixed double floats, thirty feet on either side, before the mole. These he fastened with four anchors at the four corners, that they might not be carried away by the waves. Having completed and secured them, he then joined to them other floats of equal size. These he covered over with earth and mould, that he might not be prevented from access to them to defend them, and in the front and on both sides he protected them with a parapet of wicker work; and on every fourth one raised a turret, two stories high, to secure them the better from being attacked by the shipping and set on fire.

XXVI. — To counteract this, Pompey fitted out large merchant ships, which he found in the harbour of Brundusium: on them he erected turrets three stories high, and, having furnished them with several engines and all sorts of weapons, drove them amongst Caesar's works, to break through the floats and interrupt the works; thus there happened skirmishes every day at a distance with slings, arrows, and other weapons. Caesar conducted matters as if he thought that the hopes of peace were not yet to be given up. And though he was very much surprised that Magius, whom he had sent to Pompey with a message, was not sent back to him; and though his attempting a reconciliation often retarded the vigorous prosecution of his plans, yet he thought that he ought by all means to persevere in the same line of conduct. He therefore sent Caninius Rebilus to have an interview with Scribonius Libo, his intimate friend and relation. He charges him to exhort Libo to effect a peace, but, above all things, requires that he should be admitted to an interview with Pompey. He declared that he had great hopes, if that were allowed him, that the consequence would be that both parties would lay down their arms on equal terms; that a great share of the glory and reputation of that event would redound to Libo, if, through his advice and agency, hostilities should be ended. Libo, having parted from the conference with Caninius, went to Pompey, and, shortly after, returns with answer that, as the consuls were absent, no treaty of compositions could be engaged in without them. Caesar therefore thought it time at length to give over the attempt which he had often made in vain, and act with energy in the war.

XXVII. — When Caesar's works were nearly half finished, and after nine days were spent in them, the ships which had conveyed the first division of the army to Dyrrachium being sent back by the consuls, returned to Brundusium. Pompey, either frightened at Caesar's works or determined from the beginning to quit Italy, began to prepare for his departure on the arrival of the ships; and the more effectually to retard Caesar's attack, lest his soldiers should force their way into the town at the moment of his departure, he stopped up the gates, built walls across the streets and avenues, sunk trenches across the ways, and in them fixed palisades and sharp stakes, which he made level with the ground by means of hurdles and clay. But he barricaded with large beams fastened in the ground and sharpened at the ends two passages and roads without the walls, which led to the port. After making these arrangements, he ordered his soldiers to go on board without noise, and disposed here and there, on the wall and turrets, some light-armed veterans, archers and slingers. These he designed to call off by a certain signal, when all the soldiers were embarked, and left row-galleys for them in a secure place.

XXVIII. — The people of Brundusium, irritated by the insolence of Pompey's soldiers, and the insults received from Pompey himself, were in favour of Caesar's party. Therefore, as soon as they were aware of Pompey's departure, whilst his men were running up and down, and busied about their voyage, they made signs from the tops of the houses: Caesar, being apprized of the design by them, ordered scaling ladders to be got ready, and his men to take arms, that he might not lose any opportunity of coming to an action. Pompey weighed anchor at nightfall. The soldiers who had been posted on the wall to guard it, were called off by the signal which had been agreed on, and knowing the roads, ran down to the ships. Caesar's soldiers fixed their ladders and scaled the walls: but being cautioned by the people to beware of the hidden stakes and covered trenches, they halted, and being conducted by the inhabitants by a long circuit, they reached the port, and captured with their long boats and small craft two of Pompey's ships, full of soldiers, which had struck against Caesar's moles.

XXIX. — Though Caesar highly approved of collecting a fleet, and crossing the sea, and pursuing Pompey before he could strengthen himself with his transmarine auxiliaries, with the hope of bringing the war to a conclusion, yet he dreaded the delay and length of time necessary to effect it: because Pompey, by collecting all his ships, had deprived him of the means of pursuing him at present. The only resource left to Caesar, was to wait for a fleet from the distant regions of Gaul, Picenum, and the straits of Gibraltar. But this, on account of the season of the year, appeared tedious and troublesome. He was unwilling that, in the meantime, the veteran army, and the two Spains, one of which was bound to Pompey by the strongest obligations, should be confirmed in his interest; that auxiliaries and cavalry should be provided and Gaul and Italy reduced in his absence.

XXX. — Therefore, for the present, he relinquished all intention of pursuing Pompey, and resolved to march to Spain, and commanded the magistrates of the free towns to procure him ships, and to have them conveyed to Brundusium. He detached Valerius, his lieutenant, with one legion to Sardinia; Curio, the proprietor, to Sicily with three legions; and ordered him, when he had recovered Sicily, to immediately transport his army to Africa. Marcus Cotta was at this time governor of Sardinia: Marcus Cato, of Sicily: and Tubero, by the lots, should have had the government of Africa. The Caralitani, as soon as they heard that Valerius was sent against them, even before he left Italy, of their own accord drove Cotta out of the town; who, terrified because he understood that the whole province was combined [against him], fled from Sardinia to Africa. Cato was in Sicily, repairing the old ships of war, and demanding new ones from the states, and these things he performed with great zeal. He was raising levies of Roman citizens, among the Lucani and Brutii, by his lieutenants, and exacting a certain quota of horse and foot from the states of Sicily. When these things were nearly completed, being informed of Curio's approach, he made a complaint that he was abandoned and betrayed by Pompey, who had undertaken an unnecessary war, without making any preparation, and when questioned by him and other members in the senate, had assured them that every thing was ready and provided for the war. After having made these complaints in a public assembly, he fled from his province.

XXXI. — Valerius found Sardinia, and Curio, Sicily, deserted by their governors when they arrived there with their armies. When Tubero arrived in Africa, he found Attius Varus in the government of the province, who, having lost his cohorts, as already related, at Auximum, had straightway fled to Africa, and finding it without a governor, had seized it of his own accord, and making levies, had raised two legions. From his acquaintance with the people and country, and his knowledge of that province, he found the means of effecting this; because a few years before, at the expiration of his praetorship, he had obtained that province. He, when Tubero came to Utica with his fleet, prevented his entering the port or town, and did not suffer his son, though labouring under sickness, to set foot on shore; but obliged him to weigh anchor and quit the place.

XXXIL. — When these affairs were despatched, Caesar, that there might be an intermission from labour for the rest of the season, drew off his soldiers to the nearest municipal towns, and set off in person for Rome. Having assembled the senate, he reminded them of the injustice of his enemies; and told them, "That he aimed at no extraordinary honour, but had waited for the time appointed by law, for standing candidate for the consulate, being contented with what was allowed to every citizen. That a bill had been carried by the ten tribunes of the people (notwithstanding the resistance of his enemies, and a very violent opposition from Cato, who in his usual manner, consumed the day by a tedious harangue) that he should be allowed to stand candidate, though absent, even in the consulship of Pompey; and if the latter disapproved of the bill, why did he allow it to pass? if he approved of it, why should he debar him [Caesar] from the people's favour? He made mention of his own patience, in that he had freely proposed that all armies should be disbanded, by which he himself would suffer the loss both of dignity and honour. He urged the virulence of his enemies, who refused to comply with what they required from others, and had rather that all things should be thrown into confusion, than that they should lose their power and their armies. He expatiated on their injustice, in taking away his legions: their cruelty and insolence in abridging the privileges of the tribunes; the proposals he had made, and his entreaties of an interview, which had been refused him: For which reasons, he begged and desired that they would undertake the management of the republic, and unite with him in the administration of it. But if through fear they declined it, he would not be a burden to them, but take the management of it on himself. That deputies ought to be sent to Pompey, to propose a reconciliation; as he did not regard what Pompey had lately asserted in the senate, that authority was acknowledged to be vested in those persons to whom ambassadors were sent, and fear implied in those that sent them. That these were the sentiments of low, weak minds: that for his part, as he had made it his study to surpass others in glory, so he was desirous of excelling them in justice and equity."

XXXIII. — The senate approved of sending deputies, but none could be found fit to execute the commission: for every person, from his own private fears, declined the office. For Pompey, on leaving the city, had declared in the open senate, that he would hold in the same degree of estimation, those who stayed in Rome and those in Caesar's camp. Thus three days were wasted in disputes and excuses. Besides, Lucius Metellus, one of the tribunes, was suborned by Caesar's enemies, to prevent this, and to embarrass everything else which Caesar should propose. Caesar having discovered his intention, after spending several days to no purpose, left the city, in order that he might not lose any more time, and went to Transalpine Gaul, without effecting what he had intended.

XXXIV. — On his arrival there, he was informed that, Vibullius Rufus, whom he had taken a few days before at Corfinium, and set at liberty, was sent by Pompey into Spain; and that Domitius also was gone to seize Massilia with seven row-galleys, which were fitted up by some private persons at Igilium and Cosa, and which he had manned with his own slaves, freedmen, and colonists: and that some young noblemen of Massilia had been sent before him; whom Pompey, when leaving Rome had exhorted, that the late services of Caesar should not erase from their minds the memory of his former favours. On receiving this message, the Massilians had shut their gates against Caesar, and invited over to them the Albici, who had formerly been in alliance with them, and who inhabited the mountains that overhung Massilia: they had likewise conveyed the corn from the surrounding country, and from all the forts into the city; had opened armouries in the city: and were repairing the walls, the fleet, and the gates.

XXXV. — Caesar sent for fifteen of the principal persons of Massilia to attend him. To prevent the war commencing among them, he remonstrates [in the following language]; "that they ought to follow the precedent set by all Italy, rather than submit to the will of any one man." He made use of such arguments as he thought would tend to bring them to reason. The deputies reported his speech to their countrymen, and by the authority of the state bring him back this answer: "That they understood that the Roman people were divided into two factions: that they had neither judgment nor abilities to decide which had the juster cause; but that the heads of these factions were Cneius Pompey and Caius Caesar, the two patrons of the state: the former of whom had granted to their state the lands of the Volcae Arecomici, and Helvii; the latter had assigned them a part of his conquests in Gaul, and had augmented their revenue. Wherefore, having received equal favours from both, they ought to show equal affection to both, and assist neither against the other, nor admit either into their city or harbours."

XXXVI. — Whilst this treaty was going forward, Domitius arrived at Massilia with his fleet, and was received into the city, and made governor of it. The chief management of the war was entrusted to him. At his command they send the fleet to all parts; they seize all the merchantmen they could meet with, and carry them into the harbour; they apply the nails, timber, and rigging, with which they were furnished to rig and refit their other vessels. They lay up in the public stores, all the corn that was found in the ships, and reserve the rest of their lading and convoy for the siege of the town, should such an event take place. Provoked at such ill treatment, Caesar led three legions against Massilia, and resolved to provide turrets, and vinae to assault the town, and to build twelve ships at Arelas, which being completed and rigged in thirty days (from the time the timber was cut down), and being brought to Massilia, he put under the command of Decimus Brutus; and left Caius Trebonius his lieutenant, to invest the city.

XXXVII. — Whilst he was preparing and getting these things in readiness, he sent Caius Fabius one of his lieutenants into Spain with three legions, which he had disposed in winter-quarters in Narbo, and the neighbouring country; and ordered him immediately to seize the passes of the Pyrenees, which were at that time occupied by detachments from Lucius Afranius, one of Pompey's lieutenants. He desired the other legions, which were passing the winter at a great distance, to follow close after him. Fabius, according to his orders, by using expedition, dislodged the party from the hills, and by hasty marches came up with the army of Afranius.

XXXVIII. — On the arrival of Vibullius Rufus, whom, we have already mentioned, Pompey had sent into Spain, Afranius, Petreius, and Varro, his lieutenants (one of whom had the command of Hither Spain, with three legions; the second of the country from the forest of Castulo to the river Guadiana with two legions; the third from the river Guadiana to the country of the Vettones and Lusitania, with the like number of legions), divided amongst themselves their respective departments. Petreius was to march from Lusitania through the Vettones, and join Afranius with all his forces; Varro was to guard all Further Spain with what legions he had. These matters being settled, reinforcements of horse and foot were demanded from Lusitania, by Petreius; from the Celtiberi, Cantabri, and all the barbarous nations which border on the ocean, by Afranius. When they were raised, Petreius immediately marched through the Vettones to Afranius. They resolved by joint consent to carry on the war in the vicinity of Ilerda, on account of the advantages of its situation.

XXXIX. — Afranius, as above mentioned, had three legions, Petreius two. There were besides about eighty cohorts raised in Hither and Further Spain (of which, the troops belonging to the former province had shields, those of the latter targets), and about five thousand horse raised in both provinces. Caesar had sent his legions into Spain, with about six thousand auxiliary foot, and three thousand horse, which had served under him in all his former wars, and the same number from Gaul, which he himself had provided, having expressly called out all the most noble and valiant men of each state. The bravest of these were from the Aquitani and the mountaineers, who border on the Province in Gaul. He had been informed that Pompey was marching through Mauritania with his legions to Spain, and would shortly arrive. He at the same time borrowed money from the tribunes and centurions, which he distributed amongst his soldiers. By this proceeding he gained two points; he secured the interest of the centurions by this pledge in his hands, and by his liberality he purchased the affections of his army.

XL. — Fabius sounded the inclinations of the neighbouring states by letters and messengers. He had made two bridges over the river Segre, at the distance of four miles from each other. He sent foraging parties over these bridges, because he had already consumed all the forage that was on his side of the river. The generals of Pompey's army did almost the same thing, and for the same reason: and the horse had frequent skirmishes with each other. When two of Fabius's legions had, as was their constant practice, gone forth as the usual protection to the foragers, and had crossed the river, and the baggage, and all the horse were following them, on a sudden, from the weight of the cattle, and the mass of water, the bridge fell, and all the horse were cut off from the main army, which being known to Petreius and Afranius, from the timber and hurdles that were carried down the river, Afranius immediately crossed his own bridge, which communicated between his camp and the town, with four legions and all the cavalry, and marched against Fabius's two legions. When his approach was announced, Lucius Plancus, who had the command of those legions, compelled by the emergency, took post on a rising ground; and drew up his army with two fronts, that it might not be surrounded by the cavalry. Thus, though engaged with superior numbers, he sustained the furious charge of the legions and the horse. When the battle was begun by the horse, there were observed at a distance by both sides the colours of two legions, which Caius Fabius had sent round by the further bridge to reinforce our men, suspecting, as the event verified, that the enemy's generals would take advantage of the opportunity which fortune had put in their way, to attack our men. Their approach put an end to the battle, and each general led back his legions to their respective camps.

XLI. — In two days after Caesar came to the camp with nine hundred horse, which he had retained for a bodyguard. The bridge which had been broken down by the storm was almost repaired, and he ordered it to be finished in the night. Being acquainted with the nature of the country, he left behind him six cohorts to guard the bridge, the camp, and all his baggage, and the next day set off in person for Ilerda, with all his forces drawn up in three lines, and halted just before the camp of Afranius, and having remained there a short time under arms, he offered him battle on equal terms. When this offer was made, Afranius drew out his forces, and posted them on the middle of a hill, near his camp. When Caesar perceived that Afranius declined coming to an engagement, he resolved to encamp at somewhat less than half a mile's distance from the very foot of the mountain; and that his soldiers whilst engaged in their works, might not be terrified by any sudden attack of the enemy, or disturbed in their work, he ordered them not to fortify it with a wall, which must rise high, and be seen at a distance, but draw, on the front opposite the enemy, a trench fifteen feet broad. The first and second lines continued under arms as was from the first appointed. Behind them the third line was carrying on the work without being seen; so that the whole was completed before Afranius discovered that the camp was being fortified.

XLII. — In the evening Caesar drew his legions within this trench, and rested them under arms the next night. The day following he kept his whole army within it, and as it was necessary to bring materials from a considerable distance, he for the present pursued the same plan in his work; and to each legion, one after the other, he assigned one side of the camp to fortify, and ordered trenches of the same magnitude to be cut: he kept the rest of the legions under arms without baggage to oppose the enemy. Afranius and Petreius, to frighten us and obstruct the work, drew out their forces at the very foot of the mountain, and challenged us to battle. Caesar, however, did not interrupt his work, relying on the protection of the three legions, and the strength of the fosse. After staying for a short time, and advancing no great distance from the bottom of the hill, they led back their forces to their camp. The third day Caesar fortified his camp with a rampart, and ordered the other cohorts which he had left in the upper camp, and his baggage to be removed to it.

XLIII. — Between the town of Ilerda and the next hill, on which Afranius and Petreius were encamped, there was a plain about three hundred paces broad, and near the middle of it an eminence somewhat raised above the level: Caesar hoped that if he could get possession of this and fortify it, he should be able to cut off the enemy from the town, the bridge, and all the stores which they had laid up in the town. In expectation of this he led three legions out of the camp, and, drawing up his army in an advantageous position, he ordered the advanced men of one legion to hasten forward and seize the eminence. Upon intelligence of this the cohorts which were on guard before Afranius's camp were instantly sent a nearer way to occupy the same post. The two parties engage, and as Afranius's men had reached the eminence first, our men were repulsed, and, on a reinforcement being sent, they were obliged to turn their backs and retreat to the standards of legions.

XLIV. — The manner of fighting of those soldiers was to run forward with great impetuosity and boldly take a post, and not to keep their ranks strictly, but to fight in small scattered parties: if hard pressed they thought it no disgrace to retire and give up the post, being accustomed to this manner of fighting among the Lusitanians and other barbarous nations; for it commonly happens that soldiers are strongly influenced by the customs of those countries in which they have spent much time. This method, however, alarmed our men, who were not used to such a description of warfare. For they imagined that they were about to be surrounded on their exposed flank by the single men who ran forward from their ranks; and they thought it their duty to keep their ranks, and not to quit their colours, nor, without good reason, to give up the post which they had taken. Accordingly, when the advanced guard gave way, the legion which was stationed on that wing did not keep its ground, but retreated to the next hill.

XLV. — Almost the whole army being daunted at this, because it had occurred contrary to their expectations and custom, Caesar encouraged his men and led the ninth legion to their relief, and checked the insolent and eager pursuit of the enemy, and obliged them, in their turn, to show their backs and retreat to Ilerda, and take post under the walls. But the soldiers of the ninth legion, being over zealous to repair the dishonour which had been sustained, having rashly pursued the fleeing enemy, advanced into disadvantageous ground and went up to the foot of the mountain on which the town Ilerda was built. And when they wished to retire they were again attacked by the enemy from the rising ground. The place was craggy in the front and steep on either side, and was so narrow that even three cohorts, drawn up in order of battle, would fill it; but no relief could be sent on the flanks, and the horse could be of no service to them when hard pressed. From the town, indeed, the precipice inclined with a gentle slope for near four hundred paces. Our men had to retreat this way, as they had, through their eagerness, advanced too inconsiderately. The greatest contest was in this place, which was much to the disadvantage of our troops, both on account of its narrowness, and because they were posted at the foot of the mountain, so that no weapon was thrown at them without effect: yet they exerted their valour and patience, and bore every wound. The enemy's forces were increasing, and cohorts were frequently sent to their aid from the camp through the town, that fresh men might relieve the weary. Caesar was obliged to do the same, and relieve the fatigued by sending cohorts to that post.

XLVI. — After the battle had in this manner continued incessantly for five hours, and our men had suffered much from superior numbers, having spent all their javelins, they drew their swords and charged the enemy up the hill, and, having killed a few, obliged the rest to fly. The cohorts being beaten back to the wall, and some being driven by their fears into the town, an easy retreat was afforded to our men. Our cavalry also, on either flank, though stationed on sloping or low ground, yet bravely struggled up to the top of the hill, and, riding between the two armies, made our retreat more easy and secure. Such were the various turns of fortune in the battle. In the first encounter about seventy of our men fell: amongst them Quintus Fulgenius, first centurion of the second line of the fourteenth legion, who, for his extraordinary valour, had been promoted from the lower ranks to that post. About six hundred were wounded. Of Afranius's party there were killed Titus Caecilius, principal centurion, and four other centurions, and above two hundred men.

XLVII. — But this opinion is spread abroad concerning this day, that each party thought that they came off conquerors. Afranius's soldiers, because, though they were esteemed inferior in the opinion of all, yet they had stood our attack and sustained our charge, and, at first, had kept the post and the hill which had been the occasion of the dispute; and, in the first encounter, had obliged our men to fly: but ours, because, notwithstanding the disadvantage of the ground and the disparity of numbers, they had maintained the battle for five hours, had advanced up the hill sword in hand, and had forced the enemy to fly from the higher ground and driven them into the town. The enemy fortified the hill, about which the contest had been, with strong works, and posted a garrison on it.

XLVIII. — In two days after this transaction, there happened an unexpected misfortune. For so great a storm arose, that it was agreed that there were never seen higher floods in those countries; it swept down the snow from all the mountains, and broke over the banks of the river, and in one day carried away both the bridges which Fabius had built, — a circumstance which caused great difficulties to Caesar's army. For as our camp, as already mentioned, was pitched between two rivers, the Segre and Cinca, and as neither of these could be forded for the space of thirty miles, they were all of necessity confined within these narrow limits. Neither could the states, which had espoused Caesar's cause, furnish him with corn, nor the troops, which had gone far to forage, return, as they were stopped by the waters: nor could the convoys, coming from Italy and Gaul, make their way to the camp. Besides, it was the most distressing season of the year, when there was no corn in the blade, and it was nearly ripe: and the states were exhausted, because Afranius had conveyed almost all the corn, before Caesar's arrival, into Ilerda, and whatever he had left, had been already consumed by Caesar. The cattle, which might have served as a secondary resource against want, had been removed by the states to a great distance on account of the war. They who had gone out to get forage or corn, were chased by the light troops of the Lusitanians, and the targeteers of Hither Spain, who were well acquainted with the country, and could readily swim across the river, because it is the custom of all those people not to join their armies without bladders.

XLIX. — But Afranius's army had abundance of everything; a great stock of corn had been provided and laid in long before, a large quantity was coming in from the whole province: they had a good store of forage. The bridge of Ilerda afforded an opportunity of getting all these without any danger, and the places beyond the bridge, to which Caesar had no access, were as yet untouched.

L. — Those floods continued several days. Caesar endeavoured to repair the bridges, but the height of the water did not allow him: and the cohorts disposed along the banks did not suffer them to be completed; and it was easy for them to prevent it, both from the nature of the river and the height of the water, but especially because their darts were thrown from the whole course of the bank on one confined spot; and it was no easy matter at one and the same time to execute a work in a very rapid flood, and to avoid the darts.

LI. — Intelligence was brought to Afranius that the great convoys, which were on their march to Caesar, had halted at the river. Archers from the Rutheni, and horse from the Gauls, with a long train of baggage, according to the Gallic custom of travelling, had arrived there; there were besides about six thousand people of all descriptions, with slaves and freed men. But there was no order, or regular discipline, as every one followed his own humour, and all travelled without apprehension, taking the same liberty as on former marches. There were several young noblemen, sons of senators, and of equestrian rank; there were ambassadors from several states; there were lieutenants of Caesar's. The river stopped them all. To attack them by surprise, Afranius set out in the beginning of the night, with all his cavalry and three legions, and sent the horse on before, to fall on them unawares; but the Gallic horse soon got themselves in readiness, and attacked them. Though but few, they withstood the vast number of the enemy, as long as they fought on equal terms: but when the legions began to approach, having lost a few men, they retreated to the next mountains. The delay occasioned by this battle was of great importance to the security of our men; for having gained time, they retired to the higher grounds. There were missing that day about two hundred bow-men, a few horse, and an inconsiderable number of servants and baggage.

LII. — However, by all these things, the price of provisions was raised, which is commonly a disaster attendant, not only on a time of present scarcity, but on the apprehension of future want. Provisions had now reached fifty denarii each bushel; and the want of corn had diminished the strength of the soldiers; and the inconveniences were increasing every day: and so great an alteration was wrought in a few days, and fortune had so changed sides, that our men had to struggle with the want of every necessary; while the enemy had an abundant supply of all things, and were considered to have the advantage. Caesar demanded from those states which had acceded to his alliance, a supply of cattle, as they had but little corn. He sent away the camp followers to the more distant states, and endeavoured to remedy the present scarcity by every resource in his power.

LIII. — Afranius and Petreius, and their friends, sent fuller and more circumstantial accounts of these things to Rome, to their acquaintances. Report exaggerated them so that the war appeared to be almost at an end. When these letters and despatches were received at Rome, a great concourse of people resorted to the house of Afranius, and congratulations ran high: several went out of Italy to Cneius Pompey; some of them, to be the first to bring him the intelligence; others, that they might not be thought to have waited the issue of the war, and to have come last of all.

LIV. — When Caesar's affairs were in this unfavourable position, and all the passes were guarded by the soldiers and horse of Afranius, and the bridges could not be prepared, Caesar ordered his soldiers to make ships of the kind that his knowledge of Britain a few years before had taught him. First, the keels and ribs were made of light timber, then, the rest of the hulk of the ships was wrought with wicker-work, and covered over with hides. When these were finished, he drew them down to the river in waggons in one night, a distance of twenty-two miles from his camp, and transported in them some soldiers across the river, and on a sudden took possession of a hill adjoining the bank. This he immediately fortified, before he was perceived by the enemy. To this he afterwards transported a legion: and having begun a bridge on both sides, he finished it in two days. By this means, he brought safe to his camp the convoys, and those who had gone out to forage; and began to prepare a conveyance for the provisions.

LV. — The same day he made a great part of his horse pass the river, who, falling on the foragers by surprise as they were dispersed without any suspicions, intercepted an incredible number of cattle and people; and when some Spanish light-armed cohorts were sent to reinforce the enemy, our men judiciously divided themselves into two parts, the one to protect the spoil, the other to resist the advancing foe, and to beat them back, and they cut off from the rest and surrounded one cohort, which had rashly ventured out of the line before the others, and after putting it to the sword, returned safe with considerable booty to the camp over the same bridge.

LVI. — Whilst these affairs are going forward at Ilerda, the Massilians, adopting the advice of Domitius, prepared seventeen ships of war, of which eleven were decked. To these they add several smaller vessels, that our fleet might be terrified by numbers: they man them with a great number of archers and of the Albici, of whom mention has been already made, and these they incited by rewards and promises. Domitius required certain ships for his own use, which he manned with colonists and shepherds, whom he had brought along with him. A fleet being thus furnished with every necessary, he advanced with great confidence against our ships, commanded by Decimus Brutus. It was stationed at an island opposite to Massilia.

LVII. — Brutus was much inferior in number of ships; but Caesar had appointed to that fleet the bravest men selected from all his legions, antesignani and centurions, who had requested to be employed in that service. They had provided iron hooks and harpoons, and had furnished themselves with a vast number of javelins, darts, and missiles. Thus prepared, and being apprised of the enemy's approach, they put out from the harbour, and engaged the Massilians. Both sides fought with great courage and resolution; nor did the Albici, a hardy people, bred on the highlands and inured to arms, fall much short of our men in valour: and being lately come from the Massilians, they retained in their minds their recent promises: and the wild shepherds, encouraged by the hope of liberty, were eager to prove their zeal in the presence of their masters.

LVIII. — The Massilians themselves, confiding in the quickness of their ships, and the skill of their pilots, eluded ours, and evaded the shock, and as long as they were permitted by clear space, lengthening their line they endeavoured to surround us, or to attack single ships with several of theirs, or to run across our ships, and carry away our oars, if possible; but when necessity obliged them to come nearer, they had recourse, from the skill and art of the pilots, to the valour of the mountaineers. But our men, not having such expert seamen, or skilful pilots, for they had been hastily drafted from the merchant ships, and were not yet acquainted even with the names of the rigging, were moreover impeded by the heaviness and slowness of our vessels, which having been built in a hurry and of green timber, were not so easily manoeuvred. Therefore, when Caesar's men had an opportunity of a close engagement, they cheerfully opposed two of the enemy's ships with one of theirs. And throwing in the grappling irons, and holding both ships fast, they fought on both sides of the deck, and boarded the enemy's; and having killed numbers of the Albici and shepherds, they sank some of their ships, took others with the men on board, and drove the rest into the harbour. That day the Massilians lost nine ships, including those that were taken.

LIX. — When news of this battle was brought to Caesar at Ilerda, the bridge being completed at the same time, fortune soon took a turn. The enemy, daunted by the courage of our horse, did not scour the country as freely or as boldly as before: but sometimes advancing a small distance from the camp, that they might have a ready retreat, they foraged within narrower bounds: at other times, they took a longer circuit to avoid our outposts and parties of horse; or having sustained some loss, or descried our horse at a distance, they fled in the midst of their expedition, leaving their baggage behind them; at length they resolved to leave off foraging for several days, and, contrary to the practice of all nations, to go out at night.

LX. — In the meantime the Oscenses and the Calagurritani, who were under the government of the Oscenses, send ambassadors to Caesar, and offer to submit to his orders. They are followed by the Tarraconenses, Jacetani, and Ausetani, and in a few days more by the Illurgavonenses, who dwell near the river Ebro. He requires of them all to assist him with corn, to which they agreed, and having collected all the cattle in the country, they convey them into his camp. One entire cohort of the Illurgavonenses, knowing the design of their state, came over to Caesar, from the place where they were stationed, and carried their colours with them. A great change is shortly made in the face of affairs. The bridge being finished, five powerful states being joined to Caesar, a way opened for the receiving of corn, and the rumours of the assistance of legions which were said to be on their march, with Pompey at their head, through Mauritania, having died away, several of the more distant states revolt from Afranius, and enter into league with Caesar.

LXI. — Whilst the spirits of the enemy were dismayed at these things, Caesar, that he might not be always obliged to send his horse a long circuit round by the bridge, having found a convenient place, began to sink several drains, thirty feet deep, by which he might draw off a part of the river Segre, and make a ford over it. When these were almost finished, Afranius and Petreius began to be greatly alarmed, lest they should be altogether cut off from corn and forage, because Caesar was very strong in cavalry. They therefore resolved to quit their posts, and to transfer the war to Celtiberia. There was, moreover, a circumstance that confirmed them in this resolution: for of the two adverse parties, that which had stood by Sertorius in the late war, being conquered by Pompey, still trembled at his name and sway, though absent: the other which had remained firm in Pompey's interest, loved him for the favours which they had received: but Caesar's name was not known to the barbarians. From these they expected considerable aid, both of horse and foot, and hoped to protract the war till winter, in a friendly country. Having come to this resolution, they gave orders to collect all the ships in the river Ebro, and to bring them to Octogesa, a town situated on the river Ebro, about twenty miles distant from their camp. At this part of the river, they ordered a bridge to be made of boats fastened together, and transported two legions over the river Segre, and fortified their camp with a rampart, twelve feet high.

LXII. — Notice of this being given by the scouts, Caesar continued his work day and night, with very great fatigue to the soldiers, to drain the river, and so far effected his purpose, that the horse were both able and bold enough, though with some difficulty and danger, to pass the river; but the foot had only their shoulders and upper part of their breast above the water, so that their fording it was retarded, not only by the depth of the water, but also by the rapidity of the current. However, almost at the same instant, news was received of the bridge being nearly completed over the Ebro, and a ford was found in the Segre.

LXIII. — Now indeed the enemy began to think that they ought to hasten their march. Accordingly, leaving two auxiliary cohorts in the garrison at Ilerda, they crossed the Segre with their whole force, and formed one camp with the two legions which they had led across a few days before. Caesar had no resource, but to annoy and cut down their rear; since with his cavalry to go by the bridge, required him to take a long circuit; so that they would arrive at the Ebro by a much shorter route. The horse, which he had detached, crossed the ford, and when Afranius and Petreius had broken up their camp about the third watch, they suddenly appeared on their rear, and spreading round them in great numbers, began to retard and impede their march.

LXIV. — At break of day, it was perceived from the rising grounds which joined Caesar's camp, that their rear was vigorously pressed by our horse; that the last line sometimes halted and was broken; at other times, that they joined battle and that our men were beaten back by a general charge of their cohorts, and, in their turn, pursued them when they wheeled about: but through the whole camp the soldiers gathered in parties, and declared their chagrin that the enemy had been suffered to escape from their hands and that the war had been unnecessarily protracted. They applied to their tribunes and centurions, and entreated them to inform Caesar that he need not spare their labour or consider their danger; that they were ready and able, and would venture to ford the river where the horse had crossed. Caesar, encouraged by their zeal and importunity, though he felt reluctant to expose his army to a river so exceedingly large, yet judged it prudent to attempt it and make a trial. Accordingly, he ordered all the weaker soldiers, whose spirit or strength seemed unequal to the fatigue, to be selected from each century, and left them, with one legion besides, to guard the camp: the rest of the legions he drew out without any baggage, and, having disposed a great number of horses in the river, above and below the ford, he led his army over. A few of his soldiers being carried away by the force of the current, were stopped by the horse and taken up, and not a man perished. His army being safe on the opposite bank, he drew out his forces and resolved to lead them forward in three battalions: and so great was the ardour of the soldiers that, notwithstanding the addition of a circuit of six miles and a considerable delay in fording the river, before the ninth hour of the day they came up with those who had set out at the third watch.

LXV. — When Afranius, who was in company with Petreius, saw them at a distance, being affrighted at so unexpected a sight, he halted on a rising ground and drew up his army. Caesar refreshed his army on the plain that he might not expose them to battle whilst fatigued; and when the enemy attempted to renew their march, he pursued and stopped them. They were obliged to pitch their camp sooner than they had intended, for there were mountains at a small distance; and difficult and narrow roads awaited them about five miles off. They retired behind these mountains that they might avoid Caesar's cavalry, and, placing parties in the narrow roads, stop the progress of his army and lead their own forces across the Ebro without danger or apprehension. This it was their interest to attempt and to effect by any means possible; but, fatigued by the skirmishes all day, and by the labour of their march, they deferred it till the following day: Caesar likewise encamped on the next hill.

LXVI. — About midnight a few of their men who had gone some distance from the camp to fetch water, being taken by our horse, Caesar is informed by them that the generals of the enemy were drawing their troops out of the camp without noise. Upon this information Caesar ordered the signal to be given and the military shout to be raised for packing up the baggage. When they heard the shout, being afraid lest they should be stopped in the night and obliged to engage under their baggage, or lest they should be confined in the narrow roads by Caesar's horse, they put a stop to their march and kept their forces in their camp. The next day Petreius went out privately with a few horse to reconnoitre the country. A similar movement was made from Caesar's camp. Lucius Decidius Saxa was detached with a small party to explore the nature of the country. Each returned with the same account to his camp, that there was a level road for the next five miles, that there then succeeded a rough and mountainous country. Whichever should first obtain possession of the defiles would have no trouble in preventing the other's progress.

LXVII. — There was a debate in the council between Afranius and Petreius, and the time of marching was the subject. The majority were of opinion that they should begin their march at night, "for they might reach the defiles before they should be discovered." Others, because a shout had been raised the night before in Caesar's camp, used this as an argument that they could not leave the camp unnoticed: "that Caesar's cavalry were patrolling the whole night, and that all the ways and roads were beset; that battles at night ought to be avoided, because in civil dissension, a soldier once daunted is more apt to consult his fears than his oath; that the daylight raised a strong sense of shame in the eyes of all, and that the presence of the tribunes and centurions had the same effect: by these things the soldiers would be re strained and awed to their duty. Wherefore they should, by all means, attempt to force their way by day; for, though a trifling loss might be sustained, yet the post which they desired might be secured with safety to the main body of the army." This opinion prevailed in the council, and the next day, at the dawn, they resolved to set forward.

LXVIII. — Caesar, having taken a view of the country, the moment the sky began to grow white, led his forces from the camp and marched at the head of his army by a long circuit, keeping to no regular road; for the road which led to the Ebro and Octogesa was occupied by the enemy's camp, which lay in Caesar's way. His soldiers were obliged to cross extensive and difficult valleys. Craggy cliffs, in several places, interrupted their march, insomuch that their arms had to be handed to one another, and the soldiers were forced to perform a great part of their march unarmed, and were lifted up the rocks by each other. But not a man murmured at the fatigue, because they imagined that there would be a period to all their toils if they could cut off the enemy from the Ebro and intercept their convoys.

LXIX. — At first, Afranius's soldiers ran in high spirits from their camp to look at us, and in contumelious language upbraided us, "that we were forced, for want of necessary subsistence, to run away, and return to Ilerda." For our route was different from what we proposed, and we appeared to be going a contrary way. But their generals applauded their own prudence in keeping within their camp, and it was a strong confirmation of their opinion, that they saw we marched without waggons or baggage, which made them confident that we could not long endure want. But when they saw our army gradually wheel to the right, and observed our van was already passing the line of their camp, there was nobody so stupid, or averse to fatigue, as not to think it necessary to march from the camp immediately, and oppose us. The cry to arms was raised, and all the army, except a few which were left to guard the camp, set out and marched the direct road to the Ebro.

LXX. — The contest depended entirely on despatch, which should first get possession of the defile and the mountain. The difficulty of the roads delayed Caesar's army, but his cavalry pursuing Afranius's forces, retarded their march. However, the affair was necessarily reduced to this point, with respect to Afranius's men, that if they first gained the mountains, which they desired, they would themselves avoid all danger, but could not save the baggage of their whole army, nor the cohorts which they had left behind in the camps, to which, being intercepted by Caesar's army, by no means could assistance be given. Caesar first accomplished the march, and having found a plain behind large rocks, drew up his army there in order of battle and facing the enemy. Afranius, perceiving that his rear was galled by our cavalry, and seeing the enemy before him, having come to a hill, made a halt on it. Thence he detached four cohorts of Spanish light infantry to the highest mountain which was in view: to this he ordered them to hasten with all expedition, and to take possession of it, with the intention of going to the same place with all his forces, then altering his route, and crossing the hills to Octogesa. As the Spaniards were making towards it in an oblique direction, Caesar's horse espied them and attacked them, nor were they able to withstand the charge of the cavalry even for a moment, but were all surrounded and cut to pieces in the sight of the two armies.

LXXI. — There was now an opportunity for managing affairs successfully, nor did it escape Caesar, that an army daunted at suffering such a loss before their eyes, could not stand, especially as they were surrounded by our horse, and the engagement would take place on even and open ground. To this he was importuned on all sides. The lieutenants, centurions, and tribunes, gathered round him, and begged "that he would not hesitate to begin the battle: that the hearts of all the soldiers were very anxious for it: that Afranius's men had by several circumstances betrayed signs of fear; in that they had not assisted their party; in that they had not quitted the hill; in that they did not sustain the charge of our cavalry, but crowding their standards into one place, did not observe either rank or order. But if he had any apprehensions from the disadvantage of the ground, that an opportunity would be given him of coming to battle in some other place: for that Afranius must certainly come down, and would not be able to remain there for want of water."

LXXII. — Caesar had conceived hopes of ending the affair without an engagement, or without striking a blow, because he had cut off the enemy's supplies. Why should he hazard the loss of any of his men, even in a successful battle? Why should he expose soldiers to be wounded; who had deserved so well of him? Why, in short, should he tempt fortune? especially when it was as much a general's duty to conquer by tactics, as by the sword. Besides, he was moved with compassion for those citizens, who, he foresaw, must fall: and he had rather gain his object without any loss or injury to them. This resolution of Caesar was not generally approved of; but the soldiers openly declared to each other, that since such an opportunity of victory was let pass, they would not come to an engagement, even when Caesar should wish it. He persevered however in his resolution, and retired a little from that place to abate the enemy's fears. Petreius and Afranius, having got this opportunity, retired to their camp. Caesar, having disposed parties on the mountains, and cut off all access to the Ebro, fortified his camp as close to the enemy as he could.

LXXIII. — The day following, the generals of his opponents, being alarmed that they had lost all prospect of supplies, and of access to the Ebro, consulted as to what other course they should take. There were two roads, one to Ilerda, if they chose to return, the other to Tarraco, if they should march to it. Whilst they were deliberating on these matters, intelligence was brought them that their watering parties were attacked by our horse: upon which information, they dispose several parties of horse and auxiliary foot along the road, and intermix some legionary cohorts, and begin to throw up a rampart from the camp to the water, that they might be able to procure water within their lines, both without fear, and without a guard. Petreius and Afranius divided this task between themselves, and went in person to some distance from their camp for the purpose of seeing it accomplished.

LXXIV.—The soldiers having obtained by their absence a free opportunity of conversing with each other, came out in great numbers, and inquired each for whatever acquaintance or fellow citizen he had in our camp, and invited him to him. First they returned them general thanks for sparing them the day before, when they were greatly terrified, and acknowledged that they were alive through their kindness; then they inquired about the honour of our general, and whether they could with safety entrust themselves to him; and declared their sorrow that they had not done so in the beginning, and that they had taken up arms against their relations and kinsmen. Encouraged by these conferences, they desired the general's parole for the lives of Petreius and Afranius, that they might not appear guilty of a crime, in having betrayed their generals. When they were assured of obtaining their demands, they promised that they would immediately remove their standards, and sent centurions of the first rank as deputies to treat with Caesar about a peace. In the meantime some of them invite their acquaintances, and bring them to their camp, others are brought away by their friends, so that the two camps seemed to be united into one, and several of the tribunes and centurions came to Caesar, and paid their respects to him. The same was done by some of the nobility of Spain, whom they summoned to their assistance, and kept in their camp as hostages. They inquired after their acquaintance and friends, by whom each might have the means of being recommended to Caesar. Even Afranius's son, a young man, endeavoured by means of Sulpitius the lieutenant, to make terms for his own and his father's life. Every place was filled with mirth and congratulations; in the one army, because they thought they had escaped so impending a danger; in the other, because they thought they had completed so important a matter without blows; and Caesar, in every man's judgment, reaped the advantage of his former lenity, and his conduct was applauded by all.

LXXV.—When these circumstances were announced to Afranius, he left the work which he had begun, and returned to his camp determined, as it appeared, whatever should be the event to bear it with an even and steady mind. Petreius did not neglect himself; he armed his domestics; with them and the praetorian cohort of Spaniards, and a few foreign horse, his dependants, whom he commonly kept near him to guard his person, he suddenly flew to the rampart, interrupted the conferences of the soldiers, drove our men from the camp, and put to death as many as he caught. The rest formed into a body, and, being alarmed by the unexpected danger, wrapped their left arms in their cloaks, and drew their swords, and in this manner, depending on the nearness of their camp, defended themselves against the Spaniards, and the horse, and made good their retreat to the camp, where they were protected by the cohorts, which were on guard.

LXXVI. — Petreius, after accomplishing this, went round every maniple, calling the soldiers by their names and entreating with tears, that they would not give up him and their absent general Pompey, as a sacrifice to the vengeance of their enemies. Immediately they ran in crowds to the general's pavilion, when he required them all to take an oath that they would not desert nor betray the army nor the generals, nor form any design distinct from the general interest. He himself swore first to the tenor of those words, and obliged Afranius to take the same oath. The tribunes and centurions followed their example; the soldiers were brought out by centuries, and took the same oath. They gave orders, that whoever had any of Caesar's soldiers should produce them; as soon as they were produced, they put them to death publicly in the praetorium, but most of them concealed those that they had entertained, and let them out at night over the rampart. Thus the terror raised by the generals, the cruelty of the punishments, the new obligation of an oath, removed all hopes of surrender for the present, changed the soldiers' minds, and reduced matters to the former state of war.

LXXVII. — Caesar ordered the enemy's soldiers, who had come into his camp to hold a conference, to be searched for with the strictest diligence, and sent back. But of the tribunes and centurions, several voluntarily remained with him, and he afterwards treated them with great respect. The centurions he promoted to higher ranks, and conferred on the Roman knights the honour of tribunes.

LXXVIII. — Afranius's men were distressed in foraging, and procured water with difficulty. The legionary soldiers had a tolerable supply of corn, because they had been ordered to bring from Ilerda sufficient to last twenty-two days; the Spanish and auxiliary forces had none, for they had but few opportunities of procuring any, and their bodies were not accustomed to bear burdens; and therefore a great number of them came over to Caesar every day. Their affairs were under these difficulties; but of the two schemes proposed, the most expedient seemed to be to return to Ilerda, because they had left some corn there; and there they hoped to decide on a plan for their future conduct. Tarraco lay at a greater distance; and in such a space they knew affairs might admit of many changes. Their design having met with approbation, they set out from their camp. Caesar having sent forward his cavalry, to annoy and retard their rear, followed close after with his legions. Not a moment passed in which their rear was not engaged with our horse.

LXXIX. — Their manner of fighting was this: the light cohorts closed their rear, and frequently made a stand on the level grounds. If they had a mountain to ascend, the very nature of the place readily secured them from any danger; for the advanced guards, from the rising grounds, protected the rest in their ascent. When they approached a valley or declivity, and the advanced men could not impart assistance to the tardy, our horse threw their darts at them from the rising grounds with advantage; then their affairs were in a perilous situation; the only plan left was, that whenever they came near such places, they should give orders to the legions to halt, and by a violent effort repulse our horse; and these being forced to give way, they should suddenly, with the utmost speed, run all together down to the valley, and having passed it, should face about again on the next hill. For so far were they from deriving any assistance from their horse (of which they had a large number), that they were obliged to receive them into the centre of their army, and themselves protect them, as they were daunted by former battles. And on their march no one could quit the line without being taken by Caesar's horse.

LXXX. — Whilst skirmishes were fought in this manner, they advanced but slowly and gradually, and frequently halted to help their rear, as then happened. For having advanced four miles, and being very much harassed by our horse, they took post on a high mountain, and there entrenched themselves on the front only, facing the enemy; and did not take their baggage off their cattle. When they perceived that Caesar's camp was pitched, and the tents fixed up, and his horse sent out to forage, they suddenly rushed out about twelve o'clock the same day, and, having hopes that we should be delayed by the absence of our horse, they began to march, which Caesar perceiving, followed them with the legions that remained. He left a few cohorts to guard his baggage, and ordered the foragers to be called home at the tenth hour, and the horse to follow him. The horse shortly returned to their daily duty on march, and charged the rear so vigorously, that they almost forced them to fly; and several privates and some centurions were killed. The main body of Caesar's army was at hand, and universal ruin threatened them.

LXXXI. — Then indeed, not having opportunity either to choose a convenient position for their camp, or to march forward, they were obliged to halt, and to encamp at a distance from water, and on ground naturally unfavourable. But for the reasons already given, Caesar did not attack them, nor suffer a tent to be pitched that day, that his men might be the readier to pursue them whether they attempted to run off by night or by day. Observing the defect in their position, they spent the whole night in extending their works, and turn their camp to ours. The next day, at dawn, they do the same, and spend the whole day in that manner, but in proportion as they advanced their works, and extended their camp, they were farther distant from the water; and one evil was remedied by another. The first night, no one went out for water. The next day, they left a guard in the camp, and led out all their forces to water: but not a person was sent to look for forage. Caesar was more desirous that they should be humbled by these means, and forced to come to terms, than decide the contest by battle. Yet he endeavoured to surround them with a wall and trench, that he might be able to check their most sudden sally, to which he imagined that they must have recourse. Hereupon, urged by want of fodder, that they might be the readier for a march, they killed all their baggage cattle.

LXXXII. — In this work, and the deliberations on it, two days were spent. By the third day a considerable part of Caesar's works was finished. To interrupt his progress, they drew out their legions about the eighth hour, by a certain signal, and placed them in order of battle before their camp. Caesar calling his legions off from their work, and ordering the horse to hold themselves in readiness, marshalled his army: for to appear to decline an engagement contrary to the opinion of the soldiers and the general voice, would have been attended with great disadvantage. But for the reasons already known, he was dissuaded from wishing to engage, and the more especially, because the short space between the camps, even if the enemy were put to flight, would not contribute much to a decisive victory; for the two camps were not distant from each other above two thousand feet. Two parts of this were occupied by the armies, and one third left for the soldiers to charge and make their attack. If a battle should be begun, the nearness of the camps would afford a ready retreat to the conquered party in the flight. For this reason Caesar had resolved to make resistance, if they attacked him, but not to be the first to provoke the battle.

LXXXIII. — Afranius's five legions were drawn up in two lines, the auxiliary cohorts formed the third line, and acted as reserves. Caesar had three lines, four cohorts out of each of the five legions formed the first line. Three more from each legion followed them, as reserves: and three others were behind these. The slingers and archers were stationed in the centre of the line; the cavalry closed the flanks. The hostile armies being arranged in this manner, each seemed determined to adhere to his first intention: Caesar not to hazard a battle, unless forced to it; Afranius to interrupt Caesar's works. However, the matter was deferred, and both armies kept under arms till sunset; when they both returned to their camp. The next day Caesar prepared to finish the works which he had begun. The enemy attempted to pass the river Segre by a ford. Caesar, having perceived this, sent some light-armed Germans and a party of horse across the river, and disposed several parties along the banks to guard them.

LXXXIV. — At length, beset on all sides, their cattle having been four days without fodder, and having no water, wood, or corn, they beg a conference; and that, if possible, in a place remote from the soldiers. When this was refused by Caesar, but a public interview offered if they chose it, Afranius's son was given as a hostage to Caesar. They met in the place appointed by Caesar. In the hearing of both armies, Afranius spoke thus: "That Caesar ought not to be displeased either with him or his soldiers, for wishing to preserve their attachment to their general, Cneius Pompey. That they had now sufficiently discharged their duty to him, and had suffered punishment enough, in having endured the want of every necessary: but now, pent up almost like wild beasts, they were prevented from procuring water, and prevented from walking abroad; and were not able to bear the bodily pain or the mental disgrace: but confessed themselves vanquished: and begged and entreated, if there was any room left for mercy, that they should not be necessitated to suffer the most severe penalties." These sentiments were delivered in the most submissive and humble language.

LXXXV. — Caesar replied, "That either to complain or sue for mercy became no man less than him: for that every other person had done their duty: himself, in having declined to engage on favourable terms, in an advantageous situation and time, that all things tending to a peace might be totally unembarrassed: his army, in having preserved and protected the men whom they had in their power, notwithstanding the injuries which they had received, and the murder of their comrades; and even Afranius's soldiers, who of themselves treated about concluding a peace, by which they thought that they would secure the lives of all. Thus, that the parties on both sides inclined to mercy: that the generals only were averse to peace: that they paid no regard to the laws either of conference or truce; and had most inhumanly put to death ignorant persons, who were deceived by a conference: that therefore, they had met that fate which usually befalls men from excessive obstinacy and arrogance; and were obliged to have recourse, and most earnestly desire that which they had shortly before disdained. That for his part, he would not avail himself of their present humiliation, or his present advantage, to require terms by which his power might be increased, but only that those armies, which they had maintained for so many years to oppose him, should be disbanded: for six legions had been sent into Spain, and a seventh raised there, and many and powerful fleets provided, and generals of great military experience sent to command them, for no other purpose than to oppose him; that none of these measures were adopted to keep the Spains in peace, or for the use of the province, which, from the length of the peace, stood in need of no such aid; that all these things were long since designed against him: that against him a new sort of government was established, that the same person should be at the gates of Rome, to direct the affairs of the city; and though absent, have the government of two most warlike provinces for so many years: that against him the laws of the magistrates had been altered; that the late praetors and consuls should not be sent to govern the provinces as had been the constant custom, but persons approved of and chosen by a faction. That against him the excuse of age was not admitted: but persons of tried experience in former wars were called up to take the command of the armies, that with respect to him only, the routine was not observed which had been allowed to all generals, that, after a successful war, they should return home and disband their armies, if not with some mark of honour, at least without disgrace: that he had submitted to all these things patiently, and would still submit to them: nor did he now desire to take their army from them and keep it to himself (which, however, would not be a difficult matter), but only that they should not have it to employ against him: and therefore, as he said before, let them quit the provinces, and disband their army. If this was complied with, he would injure no person; that these were the last and only conditions of peace."

LXXXVI. — It was very acceptable and agreeable to Afranius's soldiers, as might be easily known from their signs of joy, that they who expected some injury after this defeat, should obtain without solicitation the reward of a dismissal. For when a debate was introduced about the place and time of their dismissal, they all began to express, both by words and signs, from the rampart where they stood, that they should be discharged immediately: for although every security might be given that they would be disbanded, still the matter would be uncertain, if it was deferred to a future day. After a short debate on either side, it was brought to this issue: that those who had any settlement or possession in Spain, should be immediately discharged: the rest at the river Var. Caesar gave security that they should receive no damage, and that no person should be obliged against his inclination to take the military oath under him.

LXXXVII. — Caesar promised to supply them with corn from the present time, till they arrived at the river Var. He further adds, that whatever any of them lost in the war, which was in the possession of his soldiers, should be restored to those that lost them. To his soldiers he made a recompense in money for those things, a just valuation being made. Whatever disputes Afranius's soldiers had afterwards amongst themselves, they voluntarily submitted to Caesar's decision. Afranius and Petreius, when pay was demanded by the legions, a sedition almost breaking out, asserted that the time had not yet come, and required that Caesar should take cognizance of it: and both parties were content with his decision. About a third part of their army being dismissed in two days, Caesar ordered two of his legions to go before, the rest to follow the vanquished enemy: that they should encamp at a small distance from each other. The execution of this business he gave in charge to Quintus Fufius Kalenus, one of his lieutenants. According to his directions, they marched from Spain to the river Var, and there the rest of the army was disbanded.

Book II

I. — Whilst these things were going forward in Spain, Caius Trebonius, Caesar's lieutenant, who had been left to conduct the assault of Massilia, began to raise a mound, vineae, and turrets against the town, on two sides: one of which was next the harbour and docks, the other on that part where there is a passage from Gaul and Spain to that sea which forces itself up the mouth of the Rhone. For Massilia is washed almost on three sides by the sea, the remaining fourth part is the only side which has access by land. A part even of this space, which reaches to the fortress, being fortified by the nature of the country, and a very deep valley, required a long and difficult siege. To accomplish these works, Caius Trebonius sends for a great quantity of carriages and men from the whole Province, and orders hurdles and materials to be furnished. These things being provided, he raised a mound eighty feet in height.

II. — But so great a store of everything necessary for a war had been a long time before laid up in the town, and so great a number of engines, that no vineae made of hurdles could withstand their force. For poles twelve feet in length, pointed with iron, and these too shot from very large engines, sank into the ground through four rows of hurdles. Therefore the arches of the vineae were covered over with beams a foot thick, fastened together, and under this the materials of the agger were handed from one to another. Before this was carried a testudo sixty feet long, for levelling the ground, made also of very strong timber, and covered over with every thing that was capable of protecting it against the fire and stones thrown by the enemy. But the greatness of the works, the height of the wall and towers, and the multitude of engines retarded the progress of our works. Besides, frequent sallies were made from the town by the Albici, and fire was thrown on our mound and turrets. These our men easily repulsed, and, doing considerable damage to those who sallied, beat them back into the town.

III. — In the meantime, Lucius Nasidius, being sent by Cneius Pompey with a fleet of sixteen sail, a few of which had beaks of brass, to the assistance of Lucius Domitius and the Massilians, passed the straits of Sicily without the knowledge or expectation of Curio, and, putting with his fleet into Messana, and making the nobles and senate take flight with the sudden terror, carried off one of their ships out of dock. Having joined this to his other ships, he made good his voyage to Massilia, and, having sent in a galley privately, acquaints Domitius and the Massilians of his arrival, and earnestly encourages them to hazard another battle with Brutus's fleet with the addition of his aid.

IV. — The Massilians, since their former loss, had brought the same number of old ships from the docks, and had repaired and fitted them out with great industry: they had a large supply of seamen and pilots. They had got several fishing-smacks, and covered them over, that the seamen might be secure against darts: these they filled with archers and engines. With a fleet thus appointed, encouraged by the entreaties and tears of all the old men, matrons, and virgins to succour the state in this hour of distress, they went on board with no less spirit and confidence than they had fought before. For it happens, from a common infirmity of human nature, that we are more flushed with confidence, or more vehemently alarmed at things unseen, concealed, and unknown, as was the case then. For the arrival of Lucius Nasidius had filled the state with the most sanguine hopes and wishes. Having got a fair wind, they sailed out of port and went to Nasidius to Taurois, which is a fort belonging to the Massilians, and there ranged their fleet and again encouraged each other to engage, and communicated their plan of operation. The command of the right division was given to the Massilians, that of the left to Nasidius.

V. — Brutus sailed to the same place with an augmented fleet: for to those made by Caesar at Arelas were added six ships taken from the Massilians, which he had refitted since the last battle and had furnished with every necessary. Accordingly, having encouraged his men to despise a vanquished people whom they had conquered when yet unbroken, he advanced against them full of confidence and spirit. From Trebonius's camp and all the higher grounds it was easy to see into the town — how all the youth which remained in it, and all persons of more advanced years, with their wives and children, and the public guards, were either extending their hands from the wall to the heavens, or were repairing to the temples of the immortal gods, and, prostrating themselves before their images, were entreating them to grant them victory. Nor was there a single person who did not imagine that his future fortune depended on the issue of that day; for the choice of their youth and the most respectable of every age, being expressly invited and solicited, had gone on board the fleet, that if any adverse fate should befall them they might see that nothing was left for them to attempt, and, if they proved victorious, they might have hopes of preserving the city, either by their internal resources or by foreign assistance.

VI. — When the battle was begun, no effort of valour was wanting to the Massilians, but, mindful of the instructions which they had a little before received from their friends, they fought with such spirit as if they supposed that they would never have another opportunity to attempt a defence, and as if they believed that those whose lives should be endangered in the battle would not long precede the fate of the rest of the citizens, who, if the city was taken, must undergo the same fortune of war. Our ships being at some distance from each other, room was allowed both for the skill of their pilots and the manoeuvring of their ships; and if at any time ours, gaining an advantage by casting the iron hooks on board their ships, grappled with them, from all parts they assisted those who were distressed. Nor, after being joined by the Albici, did they decline coming to close engagement, nor were they much inferior to our men in valour. At the same time, showers of darts, thrown from a distance from the lesser ships, suddenly inflicted several wounds on our men when off their guard and otherwise engaged; and two of their three-decked galleys, having descried the ship of Decimus Brutus, which could be easily distinguished by its flag, rowed up against him with great violence from opposite sides: but Brutus, seeing into their designs, by the swiftness of his ship extricated himself with such address as to get clear, though only by a moment. From the velocity of their motion they struck against each other with such violence that they were both excessively injured by the shock; the beak, indeed, of one of them being broken off, the whole ship was ready to founder, which circumstance being observed, the ships of Brutus's fleet, which were nearest that station, attack them when in this disorder and sink them both.

VII. — But Nasidius's ships were of no use, and soon left the fight; for the sight of their country, or the entreaties of their relations, did not urge them to run a desperate risk of their lives. Therefore, of the number of the ships not one was lost: of the fleet of the Massilians five were sunk, four taken, and one ran off with Nasidius: all that escaped made the best of their way to Hither Spain, but one of the rest was sent forward to Massilia for the purpose of bearing this intelligence, and when it came near the city, the whole people crowded out to hear the tidings, and on being informed of the event, were so oppressed with grief, that one would have imagined that the city had been taken by an enemy at the same moment. The Massilians, however, began to make the necessary preparations for the defence of their city with unwearied energy.

VIII. — The legionary soldiers who had the management of the works on the right side observed, from the frequent sallies of the enemy, that it might prove a great protection to them to build a turret of brick under the wall for a fort and place of refuge, which they at first built low and small, [to guard them] against sudden attacks. To it they retreated, and from it they made defence if any superior force attacked them; and from it they sallied out either to repel or pursue the enemy. It extended thirty feet on every side, and the thickness of the walls was five feet. But afterwards, as experience is the best master in everything on which the wit of man is employed, it was found that it might be of considerable service if it was raised to the usual height of turrets, which was effected in the following manner.

IX. — When the turret was raised to the height for flooring, they laid it on the walls in such a manner that the ends of the joists were covered by the outer face of the wall, that nothing should project to which the enemy's fire might adhere. They, moreover, built over the joists with small bricks as high as the protection of the plutei and vineae permitted them; and on that place they laid two beams across, angle-ways, at a small distance from the outer walls, to support the rafters which were to cover the turret, and on the beams they laid joists across in a direct line, and on these they fastened down planks. These joists they made somewhat longer, to project beyond the outside of the wall, that they might serve to hang a curtain on them to defend and repel all blows whilst they were building the walls between that and the next floor, and the floor of this story they faced with bricks and mortar, that the enemy's fire might do them no damage; and on this they spread mattresses, lest the weapons thrown from engines should break through the flooring, or stones from catapults should batter the brickwork. They, moreover, made three mats of càble ropes, each of them the length of the turret walls, and four feet broad, and, hanging them round the turret on the three sides which faced the enemy, fastened them to the projecting joists. For this was the only sort of defence which, they had learned by experience in other places, could not be pierced by darts or engines. But when that part of the turret which was completed was protected and secured against every attempt of the enemy, they removed the plutei to other works. They began to suspend gradually, and raise by screws from the first-floor, the entire roof of the turret, and then they elevated it as high as the length of the mats allowed. Hid and secured within these coverings, they built up the walls with bricks, and again, by another turn of the screw, cleared a place for themselves to proceed with the building; and, when they thought it time to lay another floor, they laid the ends of the beams, covered in by the outer bricks in like manner as in the first story, and from that story they again raised the uppermost floor and the mat-work. In this manner, securely and without a blow or danger, they raised it six stories high, and in laying the materials left loop-holes in such places as they thought proper for working their engines.

X. — When they were confident that they could protect the works which lay around from this turret, they resolved to build a musculus, sixty feet long, of timber, two feet square, and to extend it from the brick tower to the enemy's tower and wall. This was the form of it: two beams of equal length were laid on the ground, at the distance of four feet from each other; and in them were fastened small pillars, five feet high, which were joined together by braces, with a gentle slope, on which the timber which they must place to support the roof of the musculus should be laid: upon this were laid beams, two feet square, bound with iron plates and nails. To the upper covering of the musculus and the upper beams, they fastened laths, four fingers square, to support the tiles which were to cover the musculus. The roof being thus sloped and laid over in rows in the same manner as the joists were laid on the braces, the musculus was covered with tiles and mortar, to secure it against fire, which might be thrown from the wall. Over the tiles hides are spread, to prevent the water let in on them by spouts from dissolving the cement of the bricks. Again, the hides were covered over with mattresses, that they might not be destroyed by fire or stones. The soldiers under the protection of the vineae, finish this whole work to the very tower, and suddenly, before the enemy were aware of it, moved it forward by naval machinery, by putting rollers under it, close up to the enemy's turret, so that it even touched the building.

XI. — The townsmen, affrighted at this unexpected stroke, bring forward with levers the largest stones they can procure; and pitching them from the wall, roll them down on the musculus. The strength of the timber withstood the shock; and whatever fell on it slid off, on account of the sloping roof. When they perceived this, they altered their plan and set fire to barrels, filled with resin and tar, and rolled them down from the wall on the musculus. As soon as they fell on it, they slid off again, and were removed from its side by long poles and forks. In the meantime, the soldiers, under cover of the musculus, were looting out with crowbars the lowest stones of the enemy's turret, with which the foundation was laid. The musculus was defended by darts, thrown from engines by our men from the brick tower, and the enemy were beaten off from the wall and turrets; nor was a fair opportunity of defending the walls given them. At length several stones being picked away from the foundation of that turret next the musculus, part of it fell down suddenly, and the rest, as if following it, leaned forward.

XII. — Hereupon, the enemy, distressed at the sudden fall of the turret, surprised at the unforeseen calamity, awed by the wrath of the gods, and dreading the pillage of their city, rush all together out of the gate unarmed, with their temples bound with fillets, and suppliantly stretch out their hands to the officers and the army. At this uncommon occurrence, the whole progress of the war was stopped, and the soldiers, turning away from the battle, ran eagerly to hear and listen to them. When the enemy came up to the commanders and the army, they all fell down at their feet, and besought them "to wait till Caesar's arrival; they saw that their city was taken, our works completed, and their tower undermined, therefore they desisted from a defence; that no obstacle could arise, to prevent their being instantly plundered at a beck, as soon as he arrived, if they refused to submit to his orders." They inform them that, "if the turret had entirely fallen down, the soldiers could not be withheld from forcing into the town and sacking it, in hopes of getting spoil." These and several other arguments to the same effect were delivered, as they were a people of great learning, with great pathos and lamentations.

XIII. — The lieutenants, moved with compassion, draw off the soldiers from the work, desist from the assault, and leave sentinels on the works. A sort of a truce having been made through compassion for the besieged, the arrival of Caesar is anxiously awaited; not a dart was thrown from the walls or by our men, but all remit their care and diligence, as if the business was at an end. For Caesar had given Trebonius strict charge not to suffer the town to be taken by storm, lest the soldiers, too much irritated both by abhorrence of their revolt, by the contempt shown to them, and by their long labour, should put to the sword all the grown-up inhabitants, as they threatened to do. And it was with difficulty that they were then restrained from breaking into the town, and they were much displeased, because they imagined that they were prevented by Trebonius from taking possession of it.

XIV. — But the enemy, destitute of all honour, only waited a time and opportunity for fraud and treachery. And after an interval of some days, when our men were careless and negligent, on a sudden, at noon, when some were dispersed, and others indulging themselves in rest on the very works, after the fatigue of the day, and their arms were all laid by and covered up, they sallied out from the gates, and, the wind being high and favourable to them, they set fire to our works; and the wind spread it in such a manner that, in the same instant, the agger, plutei, testudo, tower, and engines all caught the flames and were consumed before we could conceive how it had occurred. Our men, alarmed at such an unexpected turn of fortune, lay hold on such arms as they could find. Some rush from the camp; an attack is made on the enemy: but they were prevented, by arrows and engines from the walls, from pursuing them when they fled. They retired to their walls, and there, without fear, set the musculus and brick tower on fire. Thus, by the perfidy of the enemy and the violence of the storm, the labour of many months was destroyed in a moment. The Massilians made the same attempt the next day, having got such another storm. They sallied out against the other tower and agger, and fought with more confidence. But as our men had on the former occasion given up all thoughts of a contest, so, warned by the event of the preceding day, they had made every preparation for a defence. Accordingly, they slew several, and forced the rest to retreat into the town without effecting their design.

XV. — Trebonius began to provide and repair what had been destroyed, with much greater zeal on the part of the soldiers; for when they saw that their extraordinary pains and preparations had an unfortunate issue, they were fired with indignation that, in consequence of the impious violation of the truce, their valour should be held in derision. There was no place left them from which the materials for their mound could be fetched, in consequence of all the timber, far and wide, in the territories of the Massilians, having been cut down and carried away; they began therefore to make an agger of a new construction, never heard of before, of two walls of brick, each six feet thick, and to lay floors over them of almost the same breadth with the agger, made of timber. But wherever the space between the walls, or the weakness of the timber, seemed to require it, pillars were placed underneath and traversed beams laid on to strengthen the work, and the space which was floored was covered over with hurdles, and the hurdles plastered over with mortar. The soldiers, covered overhead by the floor, on the right and left by the wall, and in the front by the mantlets, carried whatever materials were necessary for the building without danger: the business was soon finished — the loss of their laborious work was soon repaired by the dexterity and fortitude of the soldiers. Gates for making sallies were left in the wall in such places as they thought proper.

XVI. — But when the enemy perceived that those works, which they had hoped could not be replaced without a great length of time, were put into so thorough repair by a few days' labour and diligence, that there was no room for perfidy or sallies, and that no means were left them by which they could either hurt the men by resistance or the works by fire, and when they found by former examples that their town could be surrounded with a wall and turrets on every part by which it was accessible by land, in such a manner that they could not have room to stand on their own fortifications, because our works were built almost on the top of their walls by our army, and darts could be thrown from our hands, and when they perceived that all advantage arising from their engines, on which they had built great hopes, was totally lost, and that though they had an opportunity of fighting with us on equal terms from walls and turrets, they could perceive that they were not equal to our men in bravery, they had recourse to the same proposals of surrender as before.

XVII. — In Further Spain, Marcus Varro, in the beginning of the disturbances, when he heard of the circumstances which took place in Italy, being diffident of Pompey's success, used to speak in a very friendly manner of Caesar. That though, being pre-engaged to Cneius Pompey in quality of lieutenant, he was bound in honour to him, that, nevertheless, there existed a very intimate tie between him and Caesar; that he was not ignorant of what was the duty of a lieutenant, who bore an office of trust; nor of his own strength, nor of the disposition of the whole province to Caesar. These sentiments he constantly expressed in his ordinary conversation, and did not attach himself to either party. But afterwards, when he found that Caesar was detained before Massilia, that the forces of Petreius had effected a junction with the army of Afranius, that considerable reinforcements had come to their assistance, that there were great hopes and expectations, and heard that the whole Hither province had entered into a confederacy, and of the difficulties to which Caesar was reduced afterwards at Ilerda for want of provisions, and Afranius wrote to him a fuller and more exaggerated account of these matters, he began to regulate his movements by those of fortune.

XVIII. — He made levies throughout the province; and, having completed his two legions, he added to them about thirty auxiliary cohorts: he collected a large quantity of corn to send partly to the Massilians, partly to Afranius and Petreius. He commanded the inhabitants of Gades to build ten ships of war; besides, he took care that several others should be built in Spain. He removed all the money and ornaments from the temple of Hercules to the town of Gades, and sent six cohorts thither from the province to guard them, and gave the command of the town of Gades to Caius Gallonius, a Roman knight, and friend of Domitius, who had come thither sent by Domitius to recover an estate for him; and he deposited all the arms, both public and private, in Gallonius's house. He himself [Varro] made severe harangues against Caesar. He often pronounced from his tribunal that Caesar had fought several unsuccessful battles, and that a great number of his men had deserted to Afranius. That he had these accounts from undoubted messengers, and authority on which he could rely. By these means he terrified the Roman citizens of that province, and obliged them to promise him for the service of the state one hundred and ninety thousand sesterces, twenty thousand pounds weight of silver, and a hundred and twenty thousand bushels of wheat. He laid heavier burdens on those states which he thought were friendly disposed to Caesar, and billeted troops on them; he passed judgment against some private persons, and condemned to confiscation the properties of those who had spoken or made orations against the republic, and forced the whole province to take an oath of allegiance to him and Pompey. Being informed of all that happened in Hither Spain, he prepared for war. This was his plan of operations. He was to retire with his two legions to Gades, and to lay up all the shipping and provisions there. For he had been informed that the whole province was inclined to favour Caesar's party. He thought that the war might be easily protracted in an island, if he was provided with corn and shipping. Caesar, although called back to Italy by many and important matters, yet had determined to leave no dregs of war behind him in Spain, because he knew that Pompey had many dependants and clients in the Hither province.

XIX. — Having therefore sent two legions into Further Spain under the command of Quintus Cassius, tribune of the people; he himself advances with six hundred horse by forced marches, and issues a proclamation, appointing a day on which the magistrates and nobility of all the states should attend him at Corduba. This proclamation being published through the whole province, there was not a state that did not send a part of their senate to Corduba, at the appointed time; and not a Roman citizen of any note but appeared that day. At the same time the senate at Corduba shut the gates of their own accord against Varro, and posted guards and sentinels on the wall and in the turrets, and detained two cohorts (called Colonicae, which had come there accidentally), for the defence of the town. About the same time the people of Carmona, which is by far the strongest state in the whole province, of themselves drove out of the town the cohorts, and shut the gates against them, although three cohorts had been detached by Varro to garrison the citadel.

XX. — But Varro was in greater haste on this account to reach Gades with his legion as soon as possible, lest he should be stopped either on his march or on crossing over to the island. The affection of the province to Caesar proved so great and so favourable, that he received a letter from Gades, before he was far advanced on his march: that as soon as the nobility of Gades heard of Caesar's proclamation, they had combined with the tribune of the cohorts, which were in garrison there, to drive Gallonius out of the town, and to secure the city and island for Caesar. That having agreed on the design they had sent notice to Gallonius, to quit Gades of his own accord whilst he could do it with safety; if he did not, they would take measures for themselves; that for fear of this Gallonius had been induced to quit the town. When this was known, one of Varro's two legions, which was called Vernacula, carried off the colours from Varro's camp, he himself standing by and looking on, and retired to Hispalis, and took post in the market and public places without doing any injury, and the Roman citizens residing there approved so highly of this act, that every one most earnestly offered to entertain them in their houses. When Varro, terrified at these things, having altered his route, proposed going to Italica, he was informed by his friends that the gates were shut against him. Then indeed, when intercepted from every road, he sends word to Caesar that he was ready to deliver up the legion which he commanded. He sends to him Sextus Caesar, and orders him to deliver it up to him. Varro, having delivered up the legion, went to Caesar to Corduba, and having laid before him the public accounts, handed over to him most faithfully whatever money he had, and told him what quantity of corn and shipping he had, and where.

XXI. — Caesar made a public oration at Corduba, in which he returned thanks to all severally: to the Roman citizens, because they had been zealous to keep the town in their own power; to the Spaniards, for having driven out the garrison; to the Gaditani, for having defeated the attempts of his enemies, and asserted their own liberty; to the tribunes and centurions who had gone there as a guard, for having by their valour confirmed them in their purpose. He remitted the tax which the Roman citizens had promised to Varro for the public use: he restored their goods to those who he was informed had incurred that penalty by speaking too freely, having given public and private rewards to some: he filled the rest with flattering hopes of his future intentions; and having stayed two days at Corduba, he set out for Gades: he ordered the money and ornaments which had been carried away from the temple of Hercules, and lodged in the houses of private persons, to be replaced in the temple. He made Quintus Cassius governor of the province, and assigned him four legions. He himself, with those ships which Marcus Varro had built, and others which the Gaditani had built by Varro's orders, arrived in a few days at Tarraco, where ambassadors from the greatest part of the nearer province waited his arrival. Having in the same manner conferred marks of honour both publicly and privately on some states, he left Tarraco, and went thence by land to Narbo, and thence to Massilia. There he was informed that a law was passed for creating a dictator, and that he had been nominated dictator by Marcus Lepidus the praetor.

XXII. — The Massilians, wearied out by misfortunes of every sort, reduced to the lowest ebb for want of corn, conquered in two engagements at sea, defeated in their frequent sallies, and struggling moreover with a fatal pestilence, from their long confinement and change of victuals (for they all subsisted on old millet and damaged barley, which they had formerly provided and laid up in the public stores against an emergency of this kind), their turret being demolished, a great part of their wall having given way, and despairing of any aid, either from the provinces or their armies, for these they had heard had fallen into Caesar's power, resolved to surrender now without dissimulation. But a few days before, Lucius Domitius, having discovered the intention of the Massilians, and having procured three ships, two of which he gave up to his friends, went on board the third himself, having got a brisk wind, put out to sea. Some ships, which by Brutus's orders were constantly cruising near the port, having espied him, weighed anchor, and pursued him. But of these, the ship on board of which he was, persevered itself, and continuing its flight, and by the aid of the wind got out of sight: the other two, affrighted by the approach of our galleys, put back again into the harbour. The Massilians conveyed their arms and engines out of the town, as they were ordered: brought their ships out of the port and docks, and delivered up the money in their treasury. When these affairs were despatched, Caesar, sparing the town more out of regard to their renown and antiquity than to any claim they could lay to his favour, left two legions in garrison there, sent the rest to Italy, and set out himself for Rome.

XXIII. — About the same time Caius Curio, having sailed from Sicily to Africa, and from the first despising the forces of Publius Attius Varus, transported only two of the four legions which he had received from Caesar, and five hundred horse, and having spent two days and three nights on the voyage, arrived at a place called Aquilaria, which is about twenty-two miles distant from Clupea, and in the summer season has a convenient harbour, and is enclosed by two projecting promontories. Lucius Caesar, the son, who was waiting his arrival near Clupea with ten ships which had been taken near Utica in a war with the pirates, and which Publius Attius had had repaired for this war, frightened at the number of our ships, fled the sea, and running his three-decked covered galley on the nearest shore, left her there and made his escape by land to Adrumetum. Caius Considius Longus, with a garrison of one legion, guarded this town. The rest of Caesar's fleet, after his flight, retired to Adrumetum. Marcus Rufus, the quaestor, pursued him with twelve ships, which Curio had brought from Sicily as convoy to the merchantmen, and seeing a ship left on the shore, he brought her off by a towing rope, and returned with his fleet to Curio.

XXIV. — Curio detached Marcus before with the fleet to Utica, and marched thither with his army. Having advanced two days, he came to the river Bagrada, and there left Caius Caninius Rebilus, the lieutenant, with the legions; and went forward himself with the horse to view the Cornelian camp, because that was reckoned a very eligible position for encamping. It is a straight ridge, projecting into the sea, steep and rough on both sides, but the ascent is more gentle on that part which lies opposite Utica. It is not more than a mile distant from Utica in a direct line. But on this road there is a spring, to which the sea comes up, and overflows; an extensive morass is thereby formed; and if a person would avoid it, he must make a circuit of six miles to reach the town.

XXV. — Having examined this place, Curio got a view of Varus's camp, joining the wall and town, at the gate called Bellica, well fortified by its natural situation, on one side by the town itself, on the other by a theatre which is before the town, the approaches to the town being rendered difficult and narrow by the very extensive out-buildings of that structure. At the same time he observed the roads very full of carriages and cattle which they were conveying from the country into the town on the sudden alarm. He sent his cavalry after them to plunder them and get the spoil. And at the same time Varus had detached as a guard for them six hundred Numidian horse, and four hundred foot, which king Juba had sent to Utica as auxiliaries a few days before. There was a friendship subsisting between his [Juba's] father and Pompey, and a feud between him and Curio, because he, when a tribune of the people, had proposed a law, in which he endeavoured to make public property of the kingdom of Juba. The horse engaged; but the Numidians were not able to stand our first charge; but a hundred and twenty being killed, the rest retreated into their camp near the town. In the meantime, on the arrival of his men-of-war, Curio ordered proclamation to be made to the merchant ships, which lay at anchor before Utica, in number about two hundred, that he would treat as enemies all that did not set sail immediately for the Cornelian camp. As soon as the proclamation was made, in an instant they all weighed anchor and left Utica, and repaired to the place commanded them. This circumstance furnished the army with plenty of everything.

XXVI. — After these transactions, Curio returned to his camp at Bagrada; and by a general shout of the whole army was saluted imperator. The next day he led his army to Utica, and encamped near the town. Before the works of the camp were finished, the horse upon guard brought him word that a large supply of horse and foot sent by king Juba were on their march to Utica, and at the same time a cloud of dust was observed, and in a moment the front of the line was in sight. Curio, surprised at the suddenness of the affair, sent on the horse to receive their first charge, and detain them. He immediately called off his legions from the work, and put them in battle array. The horse began the battle: and before the legions could be completely marshalled and take their ground, the king's entire forces being thrown into disorder and confusion, because they had marched without any order, and were under no apprehensions, betake themselves to flight: almost all the enemy's horse being safe, because they made a speedy retreat into the town along the shore, Caesar's soldiers slay a great number of their infantry.

XXVII. — The next night two Marsian centurions, with twenty-two men belonging to the companies, deserted from Curio's camp to Attius Varus. They, whether they uttered the sentiments which they really entertained, or wished to gratify Varus (for what we wish we readily give credit to, and what we think ourselves, we hope is the opinion of other men), assured him, that the minds of the whole army were disaffected to Curio, that it was very expedient that the armies should be brought in view of each other, and an opportunity of a conference be given. Induced by their opinion, Varus the next day led his troops out of the camp: Curio did so in like manner, and with only one small valley between them, each drew up his forces.

XXVIII. — In Varus's army there was one Sextus Quintilius Varus who, as we have mentioned before, was at Corfinium. When Caesar gave him his liberty, he went over to Africa; now, Curio had transported to Africa those legions which Caesar had received under his command a short time before at Corfinium: so that the officers and companies were still the same, excepting the change of a few centurions. Quintilius, making this a pretext for addressing them, began to go round Curio's lines, and to entreat the soldiers "not to lose all recollection of the oath which they took first to Domitius and to him their quaestor, nor bear arms against those who had shared the same fortune, and endured the same hardships in a siege, nor fight for those by whom they had been opprobriously called deserters." To this he added a few words by way of encouragement, what they might expect from his own liberality, if they should follow him and Attius. On the delivery of this speech, no intimation of their future conduct is given by Curio's army, and thus both generals led back their troops to their camp.

XXIX. — However, a great and general fear spread through Curio's camp, for it is soon increased by the various discourses of men. For every one formed an opinion of his own; and to what he had heard from others, added his own apprehensions. When this had spread from a single author to several persons, and was handed from one another, there appeared to be many authors for such sentiments as these: ["That it was a civil war; that they were men; and therefore that it was lawful for them to act freely, and follow which party they pleased." These were the legions which a short time before had belonged to the enemy; for the custom of offering free towns to those who joined the opposite party had changed Caesar's kindness. For the harshest expressions of the soldiers in general did not proceed from the Marsi and Peligni, as those which passed in the tents the night before; and some of their fellow soldiers heard them with displeasure. Some additions were also made to them by those who wished to be thought more zealous in their duty.]

XXX. — For these reasons, having called a council, Curio began to deliberate on the general welfare. There were some opinions, which advised by all means an attempt to be made, and an attack on Varus's camp; for when such sentiments prevailed among the soldiers, they thought idleness was improper. In short, they said, "that it was better bravely to try the hazard of war in a battle, than to be deserted and surrounded by their own troops, and forced to submit to the greatest cruelties." There were some who gave their opinion, that they ought to withdraw at the third watch to the Cornelian camp; that by a longer interval of time the soldiers might be brought to a proper way of thinking; and also, that if any misfortune should befall them, they might have a safer and readier retreat to Sicily, from the great number of their ships.

XXXI. — Curio, censuring both measures, said, "that the one was as deficient in spirit, as the other exceeded in it: that the latter advised a shameful flight, and the former recommended us to engage at a great disadvantage. For on what, says he, can we rely that we can storm a camp, fortified both by nature and art? Or, indeed, what advantage do we gain if we give over the assault, after having suffered considerable loss; as if success did not acquire for a general the affection of his army, and misfortune their hatred? But what does a change of camp imply but a shameful flight, and universal despair, and the alienation of the army? For neither ought the obedient to suspect that they are distrusted, nor the insolent to know that we fear them; because our fears augment the licentiousness of the latter, and diminish the zeal of the former. But if, says he, we were convinced of the truth of the reports of the disaffection of the army (which I indeed am confident are either altogether groundless, or at least less than they are supposed to be), how much better to conceal and hide our suspicions of it, than by our conduct confirm it? Ought not the defects of an army to be as carefully concealed as the wounds in our bodies, lest we should increase the enemy's hopes? but they moreover advise us to set out at midnight, in order, I suppose, that those who attempt to do wrong may have a fairer opportunity; for conduct of this kind is restrained either by shame or fear, to the display of which the night is most adverse. Wherefore, I am neither so rash as to give my opinion that we ought to attack their camp without hopes of succeeding; nor so influenced by fear as to despond: and I imagine that every expedient ought first to be tried; and I am in a great degree confident that I shall form the same opinion as yourselves on this matter."

XXXII. — Having broken up the council he called the soldiers together, and reminded them "what advantage Caesar had derived from their zeal at Corfinium; how by their good offices and influence he had brought over a great part of Italy to his interest. For, says he, all the municipal towns afterwards imitated you and your conduct; nor was it without reason that Caesar judged so favourably, and the enemy so harshly of you. For Pompey, though beaten in no engagement, yet was obliged to shift his ground, and leave Italy, from the precedent established by your conduct. Caesar committed me, whom he considered his dearest friend, and the provinces of Sicily and Africa, without which he was not able to protect Rome or Italy, to your protection. There are some here present who encourage you to revolt from us; for what can they wish for more, than at once to ruin us, and to involve you in a heinous crime? or what baser opinions could they in their resentment entertain of you, than that you would betray those who acknowledged themselves indebted to you for everything, and put yourselves in the power of those who think they have been ruined by you? Have you not heard of Caesar's exploits in Spain? that he routed two armies, conquered two generals, recovered two provinces, and effected all this within forty days after he came in sight of the enemy? Can those who were not able to stand against him whilst they were uninjured resist him when they are ruined? Will you, who took part with Caesar whilst victory was uncertain, take part with the conquered enemy when the fortune of the war is decided, and when you ought to reap the reward of your services? For they say that they have been deserted and betrayed by you, and remind you of a former oath. But did you desert Lucius Domitius, or did Lucius Domitius desert you? Did he not, when you were ready to submit to the greatest difficulties, cast you off? Did he not, without your privacy, endeavour to effect his own escape? When you were betrayed by him, were you not preserved by Caesar's generosity? And how could he think you bound by your oath to him, when, after having thrown up the ensigns of power, and abdicated his government, he became a private person, and a captive in another's power? A new obligation is left upon you, that you should disregard the oath, by which you are at present bound; and have respect only to that which was invalidated by the surrender of your general, and his diminution of rank. But I suppose, although you are pleased with Caesar, you are offended with me; however I shall not boast of my services to you, which still are inferior to my own wishes or your expectations. But, however, soldiers have ever looked for the rewards of labour at the conclusion of a war; and what the issue of it is likely to be, not even you can doubt. But why should I omit to mention my own diligence and good fortune, and to what a happy crisis affairs are now arrived? Are you sorry that I transported the army safe and entire, without the loss of a single ship? That on my arrival, in the very first attack, I routed the enemy's fleet? That twice in two days I defeated the enemy's horse? That I carried out of the very harbour and bay, two hundred of the enemy's victuallers, and reduced them to that situation that they can receive no supplies either by land or sea? Will you divorce yourselves from this fortune and these generals; and prefer the disgrace of Corfinium, the defeat of Italy, the surrender of both Spains, and the prestige of the African war? I, for my part, wished to be called a soldier of Caesar's; you honoured me with the title of Imperator. If you repent your bounty, I give it back to you; restore to me my former name that you may not appear to have conferred the honour on me as a reproach."

XXXIII. — The soldiers, being affected by this oration, frequently attempted to interrupt him whilst he was speaking, so that they appeared to bear with excessive anguish the suspicion of treachery, and when he was leaving the assembly they unanimously besought him to be of good spirits, and not hesitate to engage the enemy and put their fidelity and courage to a trial. As the wishes and opinions of all were changed by this act, Curio, with the general consent, determined, whenever opportunity offered, to hazard a battle. The next day he led out his forces and ranged them in order of battle on the same ground where they had been posted the preceding day; nor did Attius Varus hesitate to draw out his men, that, if any occasion should offer, either to tamper with our men or to engage on equal terms, he might not miss the opportunity.

XXXIV. — There lay between the two armies a valley, as already mentioned, not very deep, but of a difficult and steep ascent. Each was waiting till the enemy's forces should attempt to pass it, that they might engage with the advantage of the ground. At the same time, on the left wing, the entire cavalry of Publius Attius, and several light-armed infantry intermixed with them, were perceived descending into the valley. Against them Curio detached his cavalry and two cohorts of the Marrucini, whose first charge the enemy's horse were unable to stand, but, setting spurs to their horses, fled back to their friends: the light-infantry being deserted by those who had come out along with them, were surrounded and cut to pieces by our men. Varus's whole army, facing that way, saw their men flee and cut down. Upon which Rebilus, one of Caesar's lieutenants, whom

Curio had brought with him from Sicily knowing that he had great experience in military matters, cried out, "You see the enemy are daunted, Curio! why do you hesitate to take advantage of the opportunity?" Curio, having merely "expressed this, that the soldiers should keep in mind the professions which they had made to him the day before," then ordered them to follow him, and ran far before them all. The valley was so difficult of ascent that the foremost men could not struggle up it unless assisted by those behind. But the minds of Attius's soldiers being prepossessed with fear and the flight and slaughter of their men, never thought of opposing us; and they all imagined that they were already surrounded by our horse, and, therefore, before a dart could be thrown or our men come near them, Varus's whole army turned their backs and retreated to their camp.

XXXV.— In this flight one Fabius, a Pelignian and common soldier in Curio's army, pursuing the enemy's rear, with a loud voice shouted to Varus by his name, and often called him, so that he seemed to be one of his soldiers, who wished to speak to him and give him advice. When Varus, after being repeatedly called, stopped and looked at him, and inquired who he was and what he wanted, he made a blow with his sword at his naked shoulder and was very near killing Varus, but he escaped the danger by raising his shield to ward off the blow. Fabius was surrounded by the soldiers near him and cut to pieces; and by the multitude and crowds of those that fled, the gates of the camps were thronged and the passage stopped, and a greater number perished in that place without a stroke than in the battle and flight. Nor were we far from driving them from this camp; and some of them ran straightway to the town without halting. But both the nature of the ground and the strength of the fortifications prevented our access to the camp; for Curio's soldiers, marching out to battle, were without those things which were requisite for storming a camp. Curio, therefore, led his army back to the camp, with all his troops safe except Fabius. Of the enemy about six hundred were killed and a thousand wounded, all of whom, after Curio's return, and several more under pretext of their wounds, but in fact through fear, withdrew from the camp into the town, which Varus perceiving and knowing the terror of his army, leaving a trumpeter in his camp and a few tents for show, at the third watch led back his army quietly into the town.

XXXVI.—The next day Curio resolved to besiege Utica, and to draw lines about it. In the town there was a multitude of people, ignorant of war, owing to the length of the peace; some of them Uticans, very well inclined to Caesar, for his favours to them; the Roman population was composed of persons differing widely in their sentiments. The terror occasioned by former battles was very great; and therefore they openly talked of surrendering, and argued with Attius that he should not suffer the fortune of them all to be ruined by his obstinacy. Whilst these things were in agitation, couriers, who had been sent forward, arrived from king Juba, with the intelligence that he was on his march, with considerable forces, and encouraged them to protect and defend their city, a circumstance which greatly comforted their desponding hearts.

XXXVII. — The same intelligence was brought to Curio; but for some time he could not give credit to it, because he had so great confidence in his own good fortune. And at this time Caesar's success in Spain was announced in Africa by messages and letters. Being elated by all these things, he imagined that the king would not dare to attempt anything against him. But when he found out, from undoubted authority, that his forces were less than twenty miles distant from Utica, abandoning his works, he retired to the Cornelian camp. Here he began to lay in corn and wood, and to fortify his camp, and immediately despatched orders to Sicily, that his two legions and the remainder of his cavalry should be sent to him. His camp was well adapted for protracting a war, from the nature and strength of the situation, from its proximity to the sea, and the abundance of water and salt, of which a great quantity had been stored up from the neighbouring salt-pits. Timber could not fail him from the number of trees, nor corn, with which the lands abounded. Wherefore, with the general consent, Curio determined to wait for the rest of his forces, and protract the war.

XXXVIII. — This plan being settled, and his conduct approved of, he is informed by some deserters from the town that Juba had stayed behind in his own kingdom, being called home by a neighbouring war, and a dispute with the people of Leptis; and that Sabura, his commander-in-chief, who had been sent with a small force, was drawing near to Utica. Curio rashly believing this information, altered his design, and resolved to hazard a battle. His youth, his spirits, his former good fortune and confidence of success, contributed much to confirm this resolution. Induced by these motives, early in the night he sent all his cavalry to the enemy's camp near the river Bagrada, of which Sabura, of whom we have already spoken, was the commander. But the king was coming after them with all his forces, and was posted at a distance of six miles behind Sabura. The horse that were sent to perform their march that night, and attack the enemy unawares and unexpectedly; for the Numidians, after the usual barbarous custom, encamped here and there without any regularity. The cavalry having attacked them, when sunk in sleep and dispersed, killed a great number of them; many were frightened and ran away. After which the horse returned to Curio, and brought some prisoners with them.

XXXIX. — Curio had set out at the fourth watch with all his forces, except five cohorts which he left to guard the camp. Having advanced six miles, he met the horse, heard what had happened, and inquired from the captives who commanded the camp at Bagrada. They replied Sabura.

Through eagerness to perform his journey, he neglected to make further inquiries, but looking back to the company next him, "Don't you see, soldiers," says he, "that the answer of the prisoners corresponds with the account of the deserters, that the king is not with him, and that he sent only a small force which was not able to withstand a few horse? Hasten then to spoil, to glory; that we may now begin to think of rewarding you, and returning you thanks." The achievements of the horse were great in themselves, especially if their small number be compared with the vast host of Numidians. However, the account was enlarged by themselves, as men are naturally inclined to boast of their own merit. Besides, many spoils were produced; the men and horses that were taken were brought into their sight, that they might imagine that every moment of time which intervened was a delay to their conquest. By this means the hopes of Curio were seconded by the ardour of the soldiers. He ordered the horse to follow him, and hastened his march, that he might attack them as soon as possible, while in consternation after their flight. But the horse, fatigued by the expedition of the preceding night, were not able to keep up with him, but fell behind in different places. Even this did not abate Curio's hopes.

XL. — Juba, being informed by Sabura of the battle in the night, sent to his relief two thousand Spanish and Gallic horse, which he was accustomed to keep near him to guard his person, and that part of his infantry on which he had the greatest dependence, and he himself followed slowly after with the rest of his forces and forty elephants, suspecting that as Curio had sent his horse before, he himself would follow them. Sabura drew up his army, both horse and foot, and commanded them to give way gradually and retreat through the pretence of fear; that when it was necessary he would give them the signal for battle, and such orders as he found circumstances required. Curio, as his idea of their present behaviour was calculated to confirm his former hopes, imagined that the enemy were running away, and led his army from the rising grounds down to the plain.

XLI. — And when he had advanced from this place about sixteen miles, his army being exhausted with the fatigue, he halted. Sabura gave his men the signal, marshalled his army, and began to go around his ranks and encourage them. But he made use of the foot only for show; and sent the horse to the charge: Curio was not deficient in skill, and encouraged his men to rest all their hopes in their valour. Neither were the soldiers, though wearied, nor the horse, though few and exhausted with fatigue, deficient in ardour to engage, and courage: but the latter were in number but two hundred: the rest had dropped behind on the march. Wherever they charged they forced the enemy to give ground, but they were not able to pursue them far when they fled, or to press their horses too severely. Besides, the enemy's cavalry began to surround us on both wings and to trample down our rear. When any cohorts ran forward out of the line, the Numidians, being fresh, by their speed avoided our charge, and surrounded ours when they attempted to return to their post, and cut them off from the main body. So that it did not appear safe either to keep their ground and maintain their ranks, or to issue from the line, and run the risk. The enemy's troops were frequently reinforced by assistance sent from Juba; strength began to fail our men through fatigue; and those who had been wounded could neither quit the field nor retire to a place of safety, because the whole field was surrounded by the enemy's cavalry. Therefore, despairing of their own safety, as men usually do in the last moment of their lives, they either lamented their unhappy deaths, or recommended their parents to the survivors, if fortune should save any from the impending danger. All were full of fear and grief.

XLII. — When Curio perceived that in the general consternation neither his exhortations nor entreaties were attended to, imagining that the only hope of escaping in their deplorable situation was to gain the nearest hills, he ordered the colours to be borne that way. But a party of horse, that had been sent by Sabura, had already got possession of them. Now indeed our men were reduced to extreme despair: and some of them were killed by the cavalry in attempting to escape: some fell to the ground unhurt. Cneius Domitius, commander of the cavalry, standing round Curio with a small party of horse, urged Curio to endeavour to escape by flight, and to hasten to his camp; and assured him that he would not forsake him. But Curio declared that he would never more appear in Caesar's sight, after losing the army which had been committed by Caesar to his charge, and accordingly fought till he was killed. Very few of the horse escaped from that battle, but those who had stayed behind to refresh their horses having perceived at a distance the defeat of the whole army, retired in safety to their camp.

XLIII. — The soldiers were all killed to a man. Marcus Rufus, the quaestor, who was left behind in the camp by Curio, having got intelligence of these things, encouraged his men not to be disheartened. They beg and entreat to be transported to Sicily. He consented, and ordered the masters of the ships to have all the boats brought close to the shore early in the evening. But so great was the terror in general that some said that Juba's forces were marching up, others that Varus was hastening with his legions, and that they already saw the dust raised by their coming; of which not one circumstance had happened: others suspected that the enemy's fleet would immediately be upon them. Therefore, in the general consternation, every man consulted his own safety. Those who were on board of the fleet, were in a hurry to set sail, and their flight hastened the masters of the ships of burden. A few small fishing boats attended their duty and his orders. But as the shores were crowded, so great was the struggle to determine who of such a vast number should first get on board, that some of the vessels sank with the weight of the multitude, and the fears of the rest delayed them from coming to the shore.

XLIV. — From which circumstances it happened that a few foot and aged men, that could prevail either through interest or pity, or who were able to swim to the ships, were taken on board, and landed safe in Sicily. The rest of the troops sent their centurions as deputies to Varus at night, and surrendered themselves to him. But Juba, the next day having spied their cohorts before the town, claimed them as his booty, and ordered a great part of them to be put to the sword; a few he selected and sent home to his own realm. Although Varus complained that his honour was insulted by Juba, yet he dare not oppose him: Juba rode on horseback into the town, attended by several senators, amongst whom were Servius Sulpicius and Licinius Damasippus, and in a few days arranged and ordered what he would have done in Utica, and in a few days more returned to his own kingdom, with all his forces.

Book III

I. — Julius Caesar, holding the election as dictator, was himself appointed consul with Publius Servilius; for this was the year in which it was permitted by the laws that he should be chosen consul. This business being ended, as credit was beginning to fail in Italy, and the debts could not be paid, he determined that arbitrators should be appointed: and that they should make an estimate of the possessions and properties [of the debtors], how much they were worth before the war, and that they should be handed over in payment to the creditors. This he thought the most likely method to remove and abate the apprehension of an abolition of debt, the usual consequence of civil wars and dissensions, and to support the credit of the debtors. He likewise restored to their former condition (the praetors and tribunes first submitting the question to the people) some persons condemned for bribery at the elections, by virtue of Pompey's law, at the time when Pompey kept his legions quartered in the city (these trials were finished in a single day, one judge hearing the merits, and another pronouncing the sentences), because they had offered their service to him in the beginning of the civil war, if he chose to accept them; setting the same value on them as if he had accepted them, because they had put themselves in his power. For he had determined that they ought to be restored, rather by the judgment of the people, than appear admitted to it by his bounty: that he might neither appear ungrateful in repaying an obligation, nor arrogant in depriving the people of their prerogative of exercising this bounty.

II. — In accomplishing these things, and celebrating the Latin festival, and holding all the elections, he spent eleven days; and having resigned the dictatorship, set out from the city, and went to Brundisium, where he had ordered twelve legions and all his cavalry to meet him. But he scarcely found as many ships as would be sufficient to transport fifteen thousand legionary soldiers and five hundred horse. This [scarcity of shipping] was the only thing that prevented Caesar from putting a speedy conclusion to the war. And even these troops embarked very short of their number, because several had fallen in so many wars in Gaul, and the long march from Spain had lessened their number very much, and a severe autumn in Apulia and the district about Brundisium, after the very wholesome countries of Spain and Gaul, had impaired the health of the whole army.

III. — Pompey having got a year's respite to provide forces, during which he was not engaged in war, nor employed by an enemy, had collected a numerous fleet from Asia, and the Cyclades, from Corcyra, Athens, Pontus, Bithynia, Syria, Cilicia, Phoenicia, and Egypt, and had given directions that a great number should be built in every other place. He had exacted a large sum of money from Asia, Syria, and all the kings, dynasts, tetrarchs, and free states of Achaia; and had obliged the corporations of those provinces, of which he himself had the government, to count down to him a large sum.

IV. — He had made up nine legions of Roman citizens; five from Italy, which he had brought with him; one veteran legion from Sicily, which being composed of two, he called the Gemella; one from Crete and Macedonia, of veterans who had been discharged by their former generals, and had settled in those provinces; two from Asia, which had been levied by the activity of Lentulus. Besides he had distributed among his legions a considerable number, by way of recruits, from Thessaly, Boeotia, Achaia, and Epirus: with his legions he also intermixed the soldiers taken from Caius Antonius. Besides these, he expected two legions from Syria, with Scipio; from Crete, Lacedaemon, Pontus, Syria, and other states, he got about three thousand archers, six cohorts of slingers, two thousand mercenary soldiers, and seven thousand horse; six hundred of which, Deiotarus had brought from Gaul; Ariobarzanes, five hundred from Cappadocia. Cotus had given him about the same number from Thrace, and had sent his son Sadalis with them. From Macedonia there were two hundred, of extraordinary valour, commanded by Rascipolis; five hundred Gauls and Germans; Gabinius's troops from Alexandria, whom Aulus Gabinius had left with king Ptolemy, to guard his person. Pompey, the son, had brought in his fleet eight hundred, whom he had raised among his own and his shepherds' slaves. Tarcundarius, Castor and Donilaus had given three hundred from Gallograecia: one of these came himself, the other sent his son. Two hundred were sent from Syria by Comagenus Antiochus, whom Pompey rewarded amply. The most of them were archers. To these were added Dardanians, and Bessians, some of them mercenaries; others procured by power and influence: also, Macedonians, Thessalians, and troops from other nations and states, which completed the number which we mentioned before.

V. — He had laid in vast quantities of corn from Thessaly, Asia, Egypt, Crete, Cyrene, and other countries. He had resolved to fix his winter-quarters at Dyrrachium, Apollonia, and the other sea-ports, to hinder Caesar from passing the sea: and for this purpose had stationed his fleet along the sea-coast. The Egyptian fleet was commanded by Pompey, the son: the Asiatic, by Decimus Laelius, and Caius Triarius: the Syrian, by Caius Cassius: the Rhodian, by Caius Marcellus, in conjunction with Caius Coponius; and the Liburnian, and Achaian, by Scribonius Libo, and Marcus Octavius. But Marcus Bibulus was appointed commander-in-chief of the whole maritime department, and regulated every matter. The chief direction rested upon him.

VI. — When Caesar came to Brundisium, he made a speech to the soldiers: "That since they were now almost arrived at the termination of their toils and dangers, they should patiently submit to leave their slaves and baggage in Italy, and to embark without luggage, that a greater number of men might be put on board: that they might expect everything from victory and his liberality." They cried out with one voice, "he might give what orders he pleased, that they would cheerfully fulfil them." He accordingly set sail the fourth day of January, with seven legions on board, as already remarked. The next day he reached land, between the Ceraunian rocks and other dangerous places; meeting with a safe road for his shipping to ride in, and dreading all other ports which he imagined were in possession of the enemy, he landed his men at a place called Pharsalus, without the loss of a single vessel.

VII. — Lucretius Vespillo and Minutius Rufus were at Oricum, with eighteen Asiatic ships, which were given into their charge by the orders of Decimus Laelius: Marcus Bibulus at Corcyra, with a hundred and ten ships. But they had not the confidence to dare to move out of the harbour; though Caesar had brought only twelve ships as a convoy, only four of which had decks; nor did Bibulus, his fleet being disordered and his seamen dispersed, come up in time: for Caesar was seen at the continent before any account whatsoever of his approach had reached those regions.

VIII. — Caesar, having landed his soldiers, sent back his ships the same night to Brundisium, to transport the rest of his legions and cavalry. The charge of this business was committed to lieutenant Fufius Kalenus, with orders to be expeditious in transporting the legions. But the ships having put to sea too late, and not having taken advantage of the night breeze, fell a sacrifice on their return. For Bibulus, at Corcyra, being informed of Caesar's approach, hoped to fall in with some part of our ships, with their cargoes, but found them empty; and having taken about thirty, vented on them his rage at his own remissness, and set them all on fire: and, with the same flames, he destroyed the mariners and masters of the vessels, hoping by the severity of the punishment to deter the rest. Having accomplished this affair, he filled all the harbours and shores from Salona to Oricum with his fleets. Having disposed his guard with great care, he lay on board himself in the depth of winter, declining no fatigue or duty, and not waiting for reinforcements, in hopes that he might come within Caesar's reach.

IX. — But after the departure of the Liburnian fleet, Marcus Octavius sailed from Illyricum with what ships he had to Salona; and having spirited up the Dalmatians, and other barbarous nations, he drew Issa off from its connection with Caesar; but not being able to prevail with the council of Salona, either by promises or menaces, he resolved to storm the town. But it was well fortified by its natural situation, and a hill. The Roman citizens built wooden towers, the better to secure it; but when they were unable to resist, on account of the smallness of their numbers, being weakened by several wounds, they stooped to the last resource, and set at liberty all the slaves old enough to bear arms; and cutting the hair off the women's heads, made ropes for their engines. Octavius, being informed of their determination, surrounded the town with five encampments, and began to press them at once with a siege and storm. They were determined to endure every hardship, and their greatest distress was the want of corn. They, therefore, sent deputies to Caesar, and begged a supply from him; all other inconveniences they bore by their own resources, as well as they could: and after a long interval, when the length of the siege had made Octavius's troops more remiss than usual, having got an opportunity at noon, when the enemy were dispersed, they disposed their wives and children on the walls, to keep up the appearance of their usual attention; and forming themselves into one body, with the slaves whom they had lately enfranchised, they made an attack on Octavius's nearest camp, and having forced that, attacked the second with the same fury; and then the third and the fourth, and then the other, and beat them from them all: and having killed a great number, obliged the rest and Octavius himself to fly for refuge to their ships. This put an end to the blockade. Winter was now approaching, and Octavius, despairing of capturing the town, after sustaining such considerable losses, withdrew to Pompey, to Dyrrachium.

X.—We have mentioned that Vibullius Rufus, an officer of Pompey's, had fallen twice into Caesar's power; first at Corfinium, and afterwards in Spain. Caesar thought him a proper person, on account of his favours conferred on him, to send with proposals to Pompey: and he knew that he had an influence over Pompey. This was the substance of his proposals: "That it was the duty of both, to put an end to their obstinacy, and forbear hostilities, and not tempt fortune any further; that sufficient loss had been suffered on both sides, to serve as a lesson and instruction to them, to render them apprehensive of future calamities, by Pompey, in having been driven out of Italy, and having lost Sicily, Sardinia, and the two Spains, and one hundred and thirty cohorts of Roman citizens, in Italy and Spain: by himself, in the death of Curio, and the loss of so great an army in Africa, and the surrender of his soldiers in Corcyra. Wherefore, they should have pity on themselves, and the republic: for, from their own misfortunes, they had sufficient experience of what fortune can effect in war. That this was the only time to treat of peace; when each had confidence in his own strength, and both seemed on an equal footing. Since, if fortune showed ever so little favour to either, he who thought himself superior, would not submit to terms of accommodation; nor would he be content with an equal division, when he might expect to obtain the whole. That, as they could not agree before, the terms of peace ought to be submitted to the senate and people in Rome. That in the meantime, it ought to content the republic and themselves, if they both immediately took oath in a public assembly, that they would disband their forces within the three following days. That having divested themselves of the arms and auxiliaries, on which they placed their present confidence, they must both of necessity acquiesce in the decision of the people and senate. To give Pompey the fuller assurance of his intentions, he would dismiss all his forces on land, even his garrisons.

XI. — Vibullius, having received this commission from Caesar, thought it no less necessary to give Pompey notice of Caesar's sudden approach, that he might adopt such plans as the circumstance required, than to inform him of Caesar's message; and therefore continuing his journey by night as well as by day, and taking fresh horses for despatch, he posted away to Pompey, to inform him that Caesar was marching towards him with all his forces. Pompey was at this time in Candavia, and was on his march from Macedonia to his winter-quarters in Apollonia and Dyrrachium; but surprised at the unexpected news, he determined to go to Apollonia by speedy marches, to prevent Caesar from becoming master of all the maritime states. But as soon as Caesar had landed his troops, he set off the same day for Oricum: when he arrived there, Lucius Torquatus, who was governor of the town by Pompey's appointment, and had a garrison of Parthinians in it, endeavoured to shut the gates and defend the town, and ordered the Greeks to man the walls, and to take arms. But as they refused to fight against the power of the Roman people, and as the citizens made a spontaneous attempt to admit Caesar, despairing of any assistance, he threw open the gates, and surrendered himself and the town to Caesar, and was preserved safe from injury by him.

XII. — Having taken Oricum, Caesar marched without making any delay to Apollonia. Staberius the governor, hearing of his approach, began to bring water into the citadel, and to fortify it, and to demand hostages of the town's people. But they refuse to give any, or to shut their gates against the consul, or to take upon them to judge contrary to what all Italy and the Roman people had judged. As soon as he knew their inclinations, he made his escape privately. The inhabitants of Apollonia sent ambassadors to Caesar, and gave him admission into their town. Their example was followed by the inhabitants of Bullis, Amantia, and the other neighbouring states, and all Epirus: and they sent ambassadors to Caesar, and promised to obey his commands.

XIII. — But Pompey having received information of the transactions at Oricum and Apollonia, began to be alarmed for Dyrrachium, and endeavoured to reach it, marching day and night. As soon as it was said that Caesar was approaching, such a panic fell upon Pompey's army, because in his haste he had made no distinction between night and day, and had marched without intermission, that they almost every man deserted their colours in Epirus and the neighbouring countries; several threw down their arms, and their march had the appearance of a flight. But when Pompey had halted near Dyrrachium, and had given orders for measuring out the ground for his camp, his army even yet continuing in their fright, Labienus first stepped forward and swore that he would never desert him, and would share whatever fate fortune should assign to him. The other lieutenants took the same oath, and the tribunes and centurions followed their example: and the whole army swore in like manner. Caesar, finding the road to Dyrrachium already in the possession of Pompey, was in no great haste, but encamped by the river Apsus, in the territory of Apollonia, that the states which had deserved his support might be certain of protection from his out-guards and forts; and there he resolved to wait the arrival of his other legions from Italy, and to winter in tents. Pompey did the same; and pitching his camp on the other side of the river Apsus, collected there all his troops and auxiliaries.

XIV. — Kalenus, having put the legions and cavalry on board at Brundisium, as Caesar had directed him, as far as the number of his ships allowed, weighed anchor: and having sailed a little distance from port, received a letter from Caesar, in which he was informed, that all the ports and the whole shore was occupied by the enemy's fleet: on receiving this information he returned into the harbour, and recalled all the vessels. One of them, which continued the voyage and did not obey Kalenus's command, because it carried no troops, but was private property, bore away for Oricum, and was taken by Bibulus, who spared neither slaves nor free men, nor even children; but put all to the sword. Thus the safety of the whole army depended on a very short space of time and a great casualty.

XV. — Bibulus, as has been observed before, lay with his fleet near Oricum, and as he debarred Caesar of the liberty of the sea and harbours, so he was deprived of all intercourse with the country by land; for the whole shore was occupied by parties disposed in different places by Caesar. And he was not allowed to get either wood or water, or even anchor near the land. He was reduced to great difficulties, and distressed with extreme scarcity of every necessary; insomuch that he was obliged to bring, in transports from Corcyra, not only provisions, but even wood and water; and it once happened that, meeting with violent storms, they were forced to catch the dew by night which fell on the hides that covered their decks; yet all these difficulties they bore patiently and without repining, and thought they ought not to leave the shores and harbours free from blockade. But when they were suffering under the distress which I have mentioned, and Libo had joined Bibulus, they both called from on shipboard to Marcus Acilius and Statius Marcus, the lieutenants, one of whom commanded the town, the other the guards on the coast, that they wished to speak to Caesar on affairs of importance, if permission should be granted them. They add something further to strengthen the impression that they intended to treat about an accommodation. In the meantime they requested a truce, and obtained it from them; for what they proposed seemed to be of importance, and it was well known that Caesar desired it above all things, and it was imagined that some advantage would be derived from Bibulus's proposals.

XVI. — Caesar having set out with one legion to gain possession of the more remote states, and to provide corn, of which he had but a small quantity, was at this time at Buthrotum, opposite to Corcyra. There receiving Acilius and Marcus's letters, informing him of Libo's and Bibulus's demands, he left his legion behind him, and returned himself to Oricum. When he arrived, they were invited to a conference. Libo came and made an apology for Bibulus, "that he was a man of strong passion, and had a private quarrel against Caesar, contracted when he was aedile and praetor; that for this reason he had avoided the conference, lest affairs of the utmost importance and advantage might be impeded by the warmth of his temper. That it now was and ever had been Pompey's most earnest wish, that they should be reconciled, and lay down their arms; but they were not authorized to treat on that subject, because they resigned the whole management of the war, and all other matters, to Pompey, by order of the council. But when they were acquainted with Caesar's demands, they would transmit them to Pompey, who would conclude all of himself by their persuasions. In the meantime, let the truce be continued till the messengers could return from him; and let no injury be done on either side." To this he added a few words of the cause for which they fought, and of his own forces and resources.

XVII. — To this, Caesar did not then think proper to make any reply, nor do we now think it worth recording. But Caesar required "that he should be allowed to send commissioners to Pompey, who should suffer no personal injury; and that either they should grant it, or should take his commissioners in charge, and convey them to Pompey. That as to the truce, the war in its present state was so divided, that they by their fleet deprived him of his shipping and auxiliaries; while he prevented them from the use of the land and fresh water; and if they wished that this restraint should be removed from them, they should relinquish their blockade of the seas, but if they retained the one, he in like manner would retain the other; that nevertheless, the treaty of accommodation might still be carried on, though these points were not conceded, and that they need not be an impediment to it." They would neither receive Caesar's commissioners, nor guarantee their safety, but referred the whole to Pompey. They urged and struggled eagerly to gain the one point respecting a truce. But when Caesar perceived that they had proposed the conference merely to avoid present danger and distress, but that they offered no hopes or terms of peace, he applied his thoughts to the prosecution of the war.

XVIII. — Bibulus, being prevented from landing for several days, and being seized with a violent distemper from the cold and fatigue, as he could neither be cured on board, nor was willing to desert the charge which he had taken upon him, was unable to bear up against the violence of the disease. On his death, the sole command devolved on no single individual, but each admiral managed his own division separately, and at his own discretion. Vibullius, as soon as the alarm, which Caesar's unexpected arrival had raised, was over, began again to deliver Caesar's message in the presence of Libo, Lucius Lucceius, and Theophanes, to whom Pompey used to communicate his most confidential secrets. He had scarcely entered on the subject when Pompey interrupted him, and forbade him to proceed. "What need," says he, "have I of life or Rome, if the world shall think I enjoy them by the bounty of Caesar; an opinion which can never be removed whilst it shall be thought that I have been brought back by him to Italy, from which I set out." After the conclusion of the war, Caesar was informed of these expressions by some persons who were present at the conversation. He attempted, however, by other means to bring about a negotiation of peace.

XIX. — Between Pompey's and Caesar's camp there was only the river Apsus, and the soldiers frequently conversed with each other; and by a private arrangement among themselves, no weapons were thrown during their conferences. Caesar sent Publius Vatinius, one of his lieutenants, to the bank of the river, to make such proposals as should appear most conducive to peace; and to cry out frequently with a loud voice [asking], "Are citizens permitted to send deputies to citizens to treat of peace? a concession which had been made even to fugitives on the Pyrenean mountains, and to robbers, especially when by so doing they would prevent citizens from fighting against citizens." Having spoken much in humble language, as became a man pleading for his own and the general safety, and being listened to with silence by the soldiers of both armies, he received an answer from the enemy's party that Aulus Varro proposed coming the next day to a conference, and that deputies from both sides might come without danger, and explain their wishes, and accordingly a fixed time was appointed for the interview. When the deputies met the next day, a great multitude from both sides assembled, and the expectations of every person concerning this subject were raised very high, and their minds seemed to be eagerly disposed for peace. Titus Labienus walked forward from the crowd, and in submissive terms began to speak of peace, and to argue with Vatinius. But their conversation was suddenly interrupted by darts thrown from all sides, from which Vatinius escaped by being protected by the arms of the soldiers. However, several were wounded; and among them Cornelius Balbus, Marcus Plotius, and Lucius Tiburtius, centurions, and some privates; hereupon Labienus exclaimed, "Forbear, then, to speak any more about an accommodation, for we can have no peace unless we carry Caesar's head back with us."

XX. — At the same time in Rome, Marcus Caelius Rufus, one of the praetors, having undertaken the cause of the debtors, on entering into his office, fixed his tribunal near the bench of Caius Trebonius, the city praetor, and promised if any person appealed to him in regard to the valuation and payment of debts made by arbitration, as appointed by Caesar when in Rome, that he would relieve them. But it happened, from the justice of Trebonius's decrees and his humanity (for he thought that in such dangerous times justice should be administered with moderation and compassion), that not one could be found who would offer himself the first to lodge an appeal. For to plead poverty, to complain of his own private calamities, or the general distresses of the times, or to assert the difficulty of setting the goods to sale, is the behaviour of a man even of a moderate temper; but to retain their possessions entire, and at the same time acknowledge themselves in debt, what sort of spirit, and what impudence would it not have argued! Therefore nobody was found so unreasonable as to make such demands. But Caelius proved more severe to those very persons for whose advantage it had been designed; and starting from this beginning, in order that he might not appear to have engaged in so dishonourable an affair without effecting something, he promulgated a law, that all debts should be discharged in six equal payments, of six months each, without interest.

XXI. — When Servilius, the consul, and the other magistrates opposed him, and he himself effected less than he expected, in order to raise the passions of the people, he dropped it, and promulgated two others; one, by which he remitted the annual rents of the houses to the tenants, the other, an act of insolvency: upon which the mob made an assault on Caius Trebonius, and having wounded several persons, drove him from his tribunal. The consul Servilius informed the senate of his proceedings, who passed a decree that Caelius should be removed from the management of the republic. Upon this decree, the consul forbade him the senate; and when he was attempting to harangue the people, turned him out of the rostrum. Stung with the ignominy and with resentment, he pretended in public that he would go to Caesar, but privately sent messengers to Milo, who had murdered Clodius, and had been condemned for it; and having invited him into Italy, because he had engaged the remains of the gladiators to his interest, by making them supple presents, he joined him, and sent him to Thurinum to tamper with the shepherds. When he himself was on his road to Casilinum, at the same time that his military standards and arms were seized at Capua, his slaves seen at Naples, and the design of betraying the town discovered: his plots being revealed, and Capua shut against him, being apprehensive of danger, because the Roman citizens residing there had armed themselves, and thought he ought to be treated as an enemy to the state, he abandoned his first design, and changed his route.

XXII. — Milo in the meantime despatched letters to the free towns, purporting that he acted as he did by the orders and commands of Pompey, conveyed to him by Bibulus: and he endeavoured to engage in his interest all persons whom he imagined were under difficulties by reason of their debts. But not being able to prevail with them, he set at liberty some slaves from the work-houses, and began to assault Cosa in the district of Thurinum. There having received a blow of a stone thrown from the wall of the town which was commanded by Quintus Pedius with one legion, he died of it; and Caelius having set out, as he pretended for Caesar, went to Thurii, where he was put to death as he was tampering with some of the freemen of the town, and was offering money to Caesar's Gallic and Spanish horse, which he had sent there to strengthen the garrison. And thus these mighty beginnings, which had embroiled Italy, and kept the magistrates employed, found a speedy and happy issue.

XXIII. — Libo having sailed from Oricum, with a fleet of fifty ships, which he commanded, came to Brundisium, and seized an island, which lies opposite to the harbour; judging it better to guard that place, which was our only pass to sea, than to keep all the shores and ports blocked up by a fleet. By his sudden arrival, he fell in with some of our transports, and set them on fire, and carried off one laden with corn; he struck great terror into our men, and having in the night landed a party of soldiers and archers, he beat our guard of horse from their station, and gained so much by the advantage of situation, that he despatched letters to Pompey, that if he pleased he might order the rest of the ships to be hauled upon shore and repaired; for that with his own fleet he could prevent Caesar from receiving his auxiliaries.

XXIV. — Antonius was at this time at Brundisium, and relying on the valour of his troops, covered about sixty of the long-boats belonging to the men-of-war with penthouses and bulwarks of hurdles, and put on board them select soldiers; and disposed them separately along the shore: and under the pretext of keeping the seamen in exercise, he ordered two three-banked galleys, which he had built at Brundisium, to row to the mouth of the port. When Libo saw them advancing boldly towards him, he sent five four-banked galleys against them, in hopes of intercepting them. When these came near our ships, our veteran soldiers retreated within the harbour. The enemy, urged by their eagerness to capture them, pursued them unguardedly; for instantly the boats of Antonius, on a certain signal, rowed with great violence from all parts against the enemy; and at the first charge took one of the four-banked galleys, with the seamen and marines, and forced the rest to flee disgracefully. In addition to this loss, they were prevented from getting water by the horse which Antonius had disposed along the sea-coast. Libo, vexed at the distress and disgrace, departed from Brundisium, and abandoned the blockade.

XXV. — Several months had now elapsed, and winter was almost gone, and Caesar's legions and shipping were not coming to him from Brundisium, and he imagined that some opportunities had been neglected, for the winds had at least been often favourable, and he thought that he must trust to them at last. And the longer it was deferred, the more eager were those who commanded Pompey's fleet to guard the coast, and were more confident of preventing our getting assistance: they receive frequent reproofs from Pompey by letter, that as they had not prevented Caesar's arrival at the first, they should at least stop the remainder of his army: and they were expecting that the season for transporting troops would become more unfavourable every day, as the winds grew calmer. Caesar, feeling some trouble on this account, wrote in severe terms to his officers at Brundisium, [and gave them orders] that as soon as they found the wind to answer, they should not let the opportunity of setting sail pass by, if they were even to steer their course to the shore of Apollonia: because there they might run their ships on ground. That these parts principally were left unguarded by the enemy's fleet, because they dare not venture too far from the harbour.

XXVI. — They [his officers], exerting boldness and courage, aided by the instructions of Marcus Antonius, and Fufius Kalenus, and animated by the soldiers strongly encouraging them, and declining no danger for Caesar's safety, having got a southerly wind, weighed anchor, and the next day were carried past Apollonia and Dyrrachium, and being seen from the continent, Quintus Coponius, who commanded the Rhodian fleet at Dyrrachium, put out of the port with his ships; and when they had almost come up with us, in consequence of the breeze dying away, the south wind sprang up afresh, and rescued us. However, he did not desist from his attempt, but hoped by the labour and perseverance of his seamen to be able to bear up against the violence of the storm; and although we were carried beyond Dyrrachium, by the violence of the wind, he nevertheless continued to chase us. Our men, taking advantage of fortune's kindness, for they were still afraid of being attacked by the enemy's fleet, if the wind abated, having come near a port, called Nymphaeum, about three miles beyond Lissus, put into it (this port is protected from a south-west wind, but is not secure against a south wind); and thought less danger was to be apprehended from the storm than from the enemy. But as soon as they were within the port, the south wind, which had blown for two days, by extraordinary good luck veered round to the south-west.

XXVII. — Here one might observe the sudden turns of fortune. We who, a moment before, were alarmed for ourselves, were safely lodged in a very secure harbour: and they who had threatened ruin to our fleet, were forced to be uneasy on their own account: and thus, by a change of circumstances, the storm protected our ships, and damaged the Rhodian fleet to such a degree, that all their decked ships, sixteen in number, foundered, without exception, and were wrecked: and of the prodigious number of seamen and soldiers, some lost their lives by being dashed against the rocks, others were taken by our men: but Caesar sent them all safe home.

XXVIII. — Two of our ships, that had not kept up with the rest, being overtaken by the night, and not knowing what port the rest had made to, came to an anchor opposite Lissus. Otacilius Crassus, who commanded Pompey's fleet, detached after them several barges and small craft, and attempted to take them. At the same time, he treated with them about capitulating, and promised them their lives if they would surrender. One of them carried two hundred and twenty recruits, the other was manned with somewhat less than two hundred veterans. Here it might be seen what security men derive from a resolute spirit. For the recruits, frightened at the number of vessels, and fatigued with the rolling of the sea; and with sea-sickness, surrendered to Otacilius, after having first received his oath, that the enemy would not injure them; but as soon as they were brought before him, contrary to the obligation of his oath, they were inhumanly put to death in his presence. But the soldiers of the veteran legion, who had also struggled, not only with the inclemency of the weather, but by labouring at the pump, thought it their duty to remit nothing of their former valour: and having protracted the beginning of the night in settling the terms, under pretence of surrendering, they obliged the pilot to run the ship aground: and having got a convenient place on the shore, they spent the rest of the night there, and at daybreak, when Otacilius had sent against them a party of the horse, who guarded that part of the coast, to the number of four hundred, besides some armed men, who had followed them from the garrison, they made a brave defence, and having killed some of them, retreated in safety to our army.

XXIX. — After this action, the Roman citizens, who resided at Lissus, a town which Caesar had before assigned them, and had carefully fortified, received Antony into their town, and gave him every assistance. Otacilius, apprehensive for his own safety, escaped out of the town, and went to Pompey. All his forces, whose number amounted to three veteran legions, and one of recruits, and about eight hundred horse, being landed, Antony sent most of his ships back to Italy, to transport the remainder of the soldiers and horse. The pontons, which are a sort of Gallic ship, he left at Lissus with this object, that if Pompey, imagining Italy defenceless, should transport his army thither (and this notion was spread among the common people), Caesar might have some means of pursuing him; and he sent messengers to him with great despatch, to inform him in what part of the country he had landed his army, and what number of troops he had brought over with him.

XXX. — Caesar and Pompey received this intelligence almost at the same time; for they had seen the ships sail past Apollonia and Dyrrachium. They directed their march after them by land; but at first they were ignorant to what part they had been carried; but when they were informed of it, they each adopted a different plan; Caesar, to form a junction with Antonius as soon as possible, Pompey, to oppose Antonius's forces on their march to Caesar, and, if possible, to fall upon them unexpectedly from ambush. And the same day they both led out their armies from their winter encampment along the river Apsus; Pompey, privately by night; Caesar, openly by day. But Caesar had to march a longer circuit up the river to find a ford. Pompey's route being easy, because he was not obliged to cross the river, he advanced rapidly and by forced marches against Antonius, and being informed of his approach, chose a convenient situation, where he posted his forces; and kept his men close within camp, and forbade fires to be kindled, that his arrival might be the more secret. An account of this was immediately carried to Antonius by the Greeks. He despatched messengers to Caesar, and confined himself in his camp for one day. The next day Caesar came up with him. On learning his arrival, Pompey, to prevent his being hemmed in between two armies, quitted his position, and went with all his forces to Asparagium, in the territory of Dyrrachium, and there encamped in a convenient situation.

XXXI. — During these times, Scipio, though he had sustained some losses near mount Amanus, had assumed to himself the title of imperator, after which he demanded large sums of money from the states and princes. He had also exacted from the tax-gatherers two years' rents that they owed; and enjoined them to lend him the amount of the next year, and demanded a supply of horse from the whole province. When they were collected, leaving behind him his neighbouring enemies, the Parthians (who shortly before had killed Marcus Crassus, the imperator, and had kept Marcus Bibulus besieged), he drew his legions and cavalry out of Syria; and when he came into the province, which was under great anxiety and fear of the Parthian war, and heard some declarations of the soldiers, "That they would march against an enemy, if he would lead them on; but would never bear arms against a countryman and consul"; he drew off his legions to winter-quarters to Pergamus, and the most wealthy cities, and made them rich presents: and in order to attach them more firmly to his interest, permitted them to plunder the cities.

XXXII. — In the meantime, the money which had been demanded from the province at large, was most rigorously exacted. Besides, many new imposts of different kinds were devised to gratify his avarice. A tax of so much a head was laid on every slave and child. Columns, doors, corn, soldiers, sailors, arms, engines, and carriages, were made subject to a duty. Wherever a name could be found for anything, it was deemed a sufficient reason for levying money on it. Officers were appointed to collect it, not only in the cities, but in almost every village and fort: and whosoever of them acted with the greatest rigour and inhumanity, was esteemed the best man, and best citizen. The province was overrun with bailiffs and officers, and crowded with overseers and tax-gatherers; who, besides the duties imposed, exacted a gratuity for themselves; for they asserted, that being expelled from their own homes and countries, they stood in need of every necessary; endeavouring by a plausible pretence to colour the most infamous conduct. To this was added the most exorbitant interest, as usually happens in times of war; the whole sums being called in, on which occasion they alleged that the delay of a single day was a donation. Therefore, in those two years, the debt of the province was doubled: but notwithstanding, taxes were exacted, not only from the Roman citizens, but from every corporation and every state. And they said that these were loans, exacted by the senate's decree. The taxes of the ensuing year were demanded beforehand as a loan from the collectors, as on their first appointment.

XXXIII. — Moreover, Scipio ordered the money formerly lodged in the temple of Diana at Ephesus, to be taken out with the statues of that goddess which remained there. When Scipio came to the temple, letters were delivered to him from Pompey, in the presence of several senators, whom he had called upon to attend him; [informing him] that Caesar had crossed the sea with his legions; that Scipio should hasten to him with his army, and postpone all other business. As soon as he received the letter, he dismissed his attendants, and began to prepare for his journey to Macedonia; and a few days after set out. This circumstance saved the money at Ephesus.

XXXIV. — Caesar, having effected a junction with Antonius's army, and having drawn his legion out of Oricum, which he had left there to guard the coast, thought he ought to sound the inclination of the provinces, and march farther into the country; and when ambassadors came to him from Thessaly and Aetolia, to engage that the states in those countries would obey his orders, if he sent a garrison to protect them, he despatched Lucius Cassius Longinus, with the twenty-seventh, a legion composed of young soldiers, and two hundred horse, to Thessaly: and Caius Calvisius Sabinus, with five cohorts, and a small party of horse, into Aetolia. He recommended them to be especially careful to provide corn, because those regions were nearest to him. He ordered Cneius Domitius Calvinus to march into Macedonia with two legions, the eleventh and twelfth, and five hundred horse; from which province, Menedemus, the principal man of those regions, on that side which is called the Free, having come as ambassador, assured him of the most devoted affection of all his subjects.

XXXV. — Of these Calvisius, on his first arrival in Aetolia, being very kindly received, dislodged the enemy's garrisons in Calydon and Naupactus, and made himself master of the whole country. Cassius went to Thessaly with his legion. As there were two factions there, he found the citizens divided in their inclinations. Hegasaretus, a man of established power, favoured Pompey's interest. Petreius, a young man of a most noble family, warmly supported Caesar with his own and his friends' influence.

XXXVI. — At the same time, Domitius arrived in Macedonia: and when numerous embassies had begun to wait on him from many of the states, news was brought that Scipio was approaching with his legions, which occasioned various opinions and reports; for in strange events, rumour generally goes before. Without making any delay in any part of Macedonia, he marched with great haste against Domitius; and when he was come within about twenty miles of him, wheeled on a sudden towards Cassius Longinus in Thessaly. He effected this with such celerity, that news of his march and arrival came together; for to render his march expeditious, he left the baggage of his legions behind him at the river Haliacmon, which divides Macedonia from Thessaly, under the care of Marcus Favonius, with a guard of eight cohorts, and ordered him to build a strong fort there. At the same time, Cotus's cavalry, which used to infest the neighbourhood of Macedonia, flew to attack Cassius's camp, at which Cassius being alarmed, and having received information of Scipio's approach, and seen the horse, which he imagined to be Scipio's, he betook himself to the mountains that environ Thessaly, and thence began to make his route towards Ambracia. But when Scipio was hastening to pursue him, despatches overtook him from Favonius, that Domitius was marching against him with his legions, and that he could not maintain the garrison over which he was appointed, without Scipio's assistance. On receipt of these despatches, Scipio changed his designs and his route, desisted from his pursuit of Cassius, and hastened to relieve Favonius. Accordingly, continuing his march day and night, he came to him so opportunely, that the dust raised by Domitius's army, and Scipio's advanced guard, were observed at the same instant. Thus, the vigilance of Domitius saved Cassius, and the expedition of Scipio, Favonius.

XXXVII. — Scipio, having stayed for two days in his camp, along the river Haliacmon, which ran between him and Domitius's camp, on the third day, at dawn, led his army across a ford, and having made a regular encampment the day following, drew up his forces in front of his camp. Domitius thought he ought not to show any reluctance, but should draw out his forces and hazard a battle. But as there was a plain six miles in breadth between the two camps, he posted his army before Scipio's camp; while the latter persevered in not quitting his entrenchment. However, Domitius with difficulty restrained his men, and prevented their beginning a battle; the more so as a rivulet with steep banks, joining Scipio's camp, retarded the progress of our men. When Scipio perceived the eagerness and alacrity of our troops to engage, suspecting that he should be obliged the next day, either to fight, against his inclination, or to incur great disgrace by keeping within his camp, though he had come with high expectation, yet by advancing rashly, made a shameful end; and at night crossed the river, without even giving the signal for breaking up the camp, and returned to the ground from which he came, and there encamped near the river, on an elevated situation. After a few days, he placed a party of horse in ambush in the night, where our men had usually gone to forage for several days before. And when Quintus Varus, commander of Domitius's horse, came there as usual, they suddenly rushed from their ambush. But our men bravely supported their charge, and returned quickly every man to his own rank, and in their turn, made a general charge on the enemy: and having killed about eighty of them, and put the rest to flight, retreated to their camp with the loss of only two men.

XXXVIII. — After these transactions, Domitius, hoping to allure Scipio to a battle, pretended to be obliged to change his position through want of corn, and having given the signal for decamping, advanced about three miles, and posted his army and cavalry in a convenient place, concealed from the enemy's view. Scipio being in readiness to pursue him, detached his cavalry and a considerable number of light infantry to explore Domitius's route. When they had marched a short way, and their foremost troops were within reach of our ambush, their suspicions being raised by the neighing of the horses, they began to retreat: and the rest who followed them, observing with what speed they retreated, made a halt. Our men, perceiving that the enemy had discovered their plot, and thinking it in vain to wait for any more, having got two troops in their power, intercepted them. Among them was Marcus Opimius, general of the horse, but he made his escape: they either killed or took prisoners all the rest of these two troops, and brought them to Domitius.

XXXIX. — Caesar, having drawn his garrisons out of the sea-ports, as before mentioned, left three cohorts at Oricum to protect the town, and committed to them the charge of his ships of war, which he had transported from Italy. Acilius, as lieutenant-general, had the charge of this duty and the command of the town; he drew the ships into the inner part of the harbour, behind the town, and fastened them to the shore, and sank a merchant-ship in the mouth of the harbour to block it up; and near it he fixed another at anchor, on which he raised a turret, and faced it to the entrance of the port, and filled it with soldiers, and ordered them to keep guard against any sudden attack.

XL. — Cneius, Pompey's son, who commanded the Egyptian fleet, having got intelligence of these things, came to Oricum, and weighed up the ship, that had been sunk, with a windlass, and by straining at it with several ropes, and attacked the other which had been placed by Acilius to watch the port with several ships, on which he had raised very high turrets, so that fighting as it were from an eminence, and sending fresh men constantly to relieve the fatigued, and at the same time attempting the town on all sides by land, with ladders and his fleet, in order to divide the force of his enemies, he overpowered our men by fatigue, and the immense number of darts, and took the ship, having beat off the men who were put on board to defend it, who, however, made their escape in small boats; and at the same time he seized a natural mole on the opposite side, which almost formed an island over against the town. He carried over land, into the inner part of the harbour, four galleys, by putting rollers under them, and driving them on with levers. Then attacking on both sides the ships of war which were moored to the shore, and were not manned, he carried off four of them, and set the rest on fire. After despatching this business, he left Decimus Laelius, whom he had taken away from the command of the Asiatic fleet, to hinder provisions from being brought into the town from Biblis and Amantia, and went himself to Lissus, where he attacked thirty merchantmen, left within the port by Antonius, and set them on fire. He attempted to storm Lissus, but being delayed three days by the vigorous defence of the Roman citizens who belonged to that district, and of the soldiers which Caesar had sent to keep garrison there, and having lost a few men in the assault, he returned without effecting his object.

XLI. — As soon as Caesar heard that Pompey was at Asparagium, he set out for that place with his army, and having taken the capital of the Parthinians on his march, where there was a garrison of Pompey's, he reached Pompey in Macedonia, on the third day, and encamped beside him; and the day following, having drawn out all his forces before his camp, he offered Pompey battle. But perceiving that he kept within his trenches, he led his army back to his camp, and thought of pursuing some other plan. Accordingly, the day following, he set out with all his forces by a long circuit, through a difficult and narrow road to Dyrrachium; hoping, either that Pompey would be compelled to follow him to Dyrrachium, or that his communication with it might be cut off, because he had deposited there all his provisions and materiel of war. And so it happened; for Pompey, at first not knowing his design, because he imagined he had taken a route in a different direction from that country, thought that the scarcity of provisions had obliged him to shift his quarters; but having afterwards got true intelligence from his scouts, he decamped the day following, hoping to prevent him by taking a shorter road; which Caesar suspecting might happen, encouraged his troops to submit cheerfully to the fatigue, and having halted a very small part of the night, he arrived early in the morning at Dyrrachium, when the van of Pompey's army was visible at a distance, and there he encamped.

XLII. — Pompey, being cut off from Dyrrachium, as he was unable to effect his purpose, took a new resolution, and entrenched himself strongly on a rising ground, which is called Petra, where ships of a small size can come in, and be sheltered from some winds. Here he ordered a part of his men-of-war to attend him, and corn and provisions to be brought from Asia, and from all the countries of which he kept possession. Caesar, imagining that the war would be protracted to too great a length, and despairing of his convoys from Italy, because all the coasts were guarded with great diligence by Pompey's adherents; and because his own fleets, which he had built during the winter, in Sicily, Gaul, and Italy, were detained; sent Lucius Canuleius into Epirus to procure corn; and because these countries were too remote, he fixed granaries in certain places, and regulated the carriage of the corn for the neighbouring states. He likewise gave directions that search should be made for whatever corn was in Lissus, the country of the Parthini, and all the places of strength. The quantity was very small, both from the nature of the land (for the country is rough and mountainous, and the people commonly import what grain they use); and because Pompey had foreseen what would happen, and some days before had plundered the Parthini, and having ravaged and dug up their houses, carried off all the corn, which he collected by means of his horse.

XLIII. — Caesar, on being informed of these transactions, pursued measures suggested by the nature of the country. For round Pompey's camps there were several high and rough hills. These he first of all occupied with guards, and raised strong forts on them. Then drawing a fortification from one fort to another, as the nature of each position allowed, he began to draw a line of circumvallation round Pompey; with these views; as he had but a small quantity of corn, and Pompey was strong in cavalry, that he might furnish his army with corn and other necessaries from all sides with less danger: secondly, to prevent Pompey from foraging, and thereby render his horse ineffectual in the operations of the war; and thirdly, to lessen his reputation, on which he saw he depended greatly, among foreign nations, when a report should have spread throughout the world that he was blockaded by Caesar, and dare not hazard a battle.

XLIV. — Neither was Pompey willing to leave the sea and Dyrrachium, because he had lodged his materiel there, his weapons, arms, and engines; and supplied his army with corn from it by his ships: nor was he able to put a stop to Caesar's works without hazarding a battle, which at that time he had determined not to do. Nothing was left but to adopt the last resource, namely, to possess himself of as many hills as he could, and cover as great an extent of country as possible with his troops, and divide

Caesar's forces as much as possible; and so it happened: for having raised twenty-four forts, and taken in a compass of fifteen miles, he got forage in this space, and within this circuit there were several fields lately sown, in which the cattle might feed in the meantime. And as our men, who had completed their works by drawing lines of communication from one fort to another, were afraid that Pompey's men would sally out from some part, and attack us in the rear; so the enemy were making a continued fortification in a circuit within ours to prevent us from breaking in on any side, or surrounding them on the rear. But they completed their works first; both because they had a greater number of men, and because they had a smaller compass to enclose. When Caesar attempted to gain any place, though Pompey had resolved not to oppose him with his whole force or to come to a general engagement; yet he detached to particular places slingers and archers, with which his army abounded, and several of our men were wounded, and filled with great dread of the arrows; and almost all the soldiers made coats or coverings for themselves of hair cloths, tarpaulins, or raw hides to defend them against the weapons.

XLV. — In seizing the posts, each exerted his utmost power: Caesar, to confine Pompey within as narrow a compass as possible; Pompey, to occupy as many hills as he could in as large a circuit as possible, and several skirmishes were fought in consequence of it. In one of these, when Caesar's ninth legion had gained a certain post, and had begun to fortify it; Pompey possessed himself of a hill near to and opposite the same place, and endeavoured to annoy the men while at work; and as the approach on one side was almost level, he first surrounded it with archers and slingers, and afterwards by detaching a strong party of light infantry, and using his engines, he stopped our works: and it was no easy matter for our men at once to defend themselves, and to proceed with their fortifications. When Caesar perceived that his troops were wounded from all sides, he determined to retreat and give up the post; his retreat was down a precipice, on which account they pushed on with more spirit, and would not allow us to retire, because they imagined that we resigned the place through fear. It is reported that Pompey said that day in triumph to his friends about him, "That he would consent to be accounted a general of no experience, if Caesar's legions effected a retreat without considerable loss from that ground into which they had rashly advanced."

XLVI. — Caesar, being uneasy about the retreat of his soldiers, ordered hurdles to be carried to the further side of the hill, and to be placed opposite to the enemy, and behind them a trench of a moderate breadth to be sunk by his soldiers under shelter of the hurdles: and the ground to be made as difficult as possible. He himself disposed slingers in convenient places to cover our men in their retreat. These things being completed, he ordered his legions to file off. Pompey's men insultingly and boldly pursued and chased us, levelling the hurdles that were thrown up in the front of our works, in order to pass over the trench. Which as soon as Caesar perceived, being afraid that his men would appear not to retreat, but to be repulsed, and that greater loss might be sustained, when his men were almost half way down the hill, he encouraged them by Antonius, who commanded that legion, ordered the signal of battle to be sounded, and a charge to be made on the enemy. The soldiers of the ninth legion suddenly closing their files threw their javelins, and advancing impetuously from the low ground up the steep, drove Pompey's men precipitately before them, and obliged them to turn their backs; but their retreat was greatly impeded by the hurdles that lay in a long line before them, and the pallisades which were in their way, and the trenches that were sunk. But our men being contented to retreat without injury, having killed several of the enemy, and lost but five of their own, very quietly retired, and having seized some other hills somewhat on this side of that place, completed their fortifications.

XLVII. — This method of conducting a war was new and unusual, as well on account of the number of forts, the extent and greatness of the works, and the manner of attack and defence, as on account of other circumstances. For all who have attempted to besiege any person, have attacked the enemy when they were frightened or weak, or after a defeat; or have been kept in fear of some attack, when they themselves have had a superior force both of foot and horse. Besides, the usual design of a siege is to cut off the enemy's supplies. On the contrary, Caesar, with an inferior force, was enclosing troops sound and unhurt, and who had abundance of all things. For there arrived every day a prodigious number of ships, which brought them provisions: nor could the wind blow from any point that would not be favourable to some of them. Whereas, Caesar, having consumed all the corn far and near, was in very great distress, but his soldiers bore all with uncommon patience. For they remembered that they lay under the same difficulties last year in Spain, and yet by labour and patience had concluded a dangerous war. They recollected too that they had suffered an alarming scarcity at Alesia, and a much greater at Avaricum, and yet had returned victorious over mighty nations. They refused neither barley nor pulse when offered them, and they held in great esteem cattle, of which they got great quantities from Epirus.

XLVIII. — There was a sort of root, called chara, discovered by the troops which served under Valerius. This they mixed up with milk, and it greatly contributed to relieve their want. They made it into a sort of bread. They had great plenty of it: loaves made of this, when Pompey's men upbraided ours with want, they frequently threw among them to damp their hopes.

XLIX. — The corn was now beginning to ripen, and their hope supported their want, as they were confident of having abundance in a short time. And there were frequently heard declarations of the soldiers on guard, in discourse with each other, that they would rather live on the bark of the trees, than let Pompey escape from their hands. For they were often told by deserters, that they could scarcely maintain their horses, and that their other cattle were dead: that they themselves were not in good health from their confinement within so narrow a compass, from the noisome smell, the number of carcasses, and the constant fatigue to them, being men unaccustomed to work, and labouring under a great want of water. For Caesar had either turned the course of all the rivers and streams which ran to the sea, or had dammed them up with strong works. And as the country was mountainous, and the valleys narrow at the bottom, he enclosed them with piles sunk in the ground, and heaped up mould against them to keep in the water. They were therefore obliged to search for low and marshy grounds, and to sink wells, and they had this labour in addition to their daily works. And even these springs were at a considerable distance from some of their posts, and soon dried up with the heat. But Caesar's army enjoyed perfect health and abundance of water, and had plenty of all sorts of provisions except corn; and they had a prospect of better times approaching, and saw greater hopes laid before them by the ripening of the grain.

L. — In this new kind of war, new methods of managing it were invented by both generals. Pompey's men, perceiving by our fires at night, at what part of the works our cohorts were on guard, coming silently upon them discharged their arrows at random among the whole multitude, and instantly retired to their camp: as a remedy against which our men were taught by experience to light their fires in one place, and keep guard in another.

* * * * *

LI. — In the meantime, Publius Sylla, whom Caesar at his departure had left governor of his camp, came up with two legions to assist the cohort; upon whose arrival Pompey's forces were easily repulsed. Nor did they stand the sight and charge of our men, and the foremost falling, the rest turned their backs and quitted the field. But Sylla called our men in from the pursuit, lest their ardour should carry them too far, but most people imagine, that if he had consented to a vigorous pursuit, the war might have been ended that day. His conduct however does not appear to deserve censure; for the duties of a lieutenant-general and of a commander-in-chief are very different; the one is bound to act entirely according to his instructions, the other to regulate his conduct without control, as occasion requires. Sylla, being deputed by Caesar to take care of the camp, and having rescued his men, was satisfied with that, and did not desire to hazard a battle (although this circumstance might probably have had a successful issue), that he might not be thought to have assumed the part of the general. One circumstance laid the Pompeians under great difficulty in making good a retreat: for they had advanced from disadvantageous ground, and were posted on the top of a hill. If they attempted to retire down the steep, they dreaded the pursuit of our men from the rising ground, and there was but a short time till sunset: for in hopes of completing the business, they had protracted the battle almost till night. Taking therefore measures suited to their exigency, and to the shortness of the time, Pompey possessed himself of an eminence, at such a distance from our fort, that no weapon discharged from an engine could reach him. Here he took up a position, and fortified it, and kept all his forces there.

LII. — At the same time, there were engagements in two other places; for Pompey had attacked several forts at once, in order to divide our forces; that no relief might be sent from the neighbouring posts. In one place, Volcatius Tullus sustained the charge of a legion with three cohorts, and beat them off the field. In another, the Germans, having sallied over our fortifications, slew several of the enemy, and retreated safe to our camp.

LIII. — Thus six engagements having happened in one day, three at Dyrrachium, and three at the fortifications, when a computation was made of the number of slain, we found that about two thousand fell on Pompey's side, several of them volunteer veterans and centurions. Among them was Valerius, the son of Lucius Flaccus, who as praetor had formerly had the government of Asia, and six military standards were taken. Of our men, not more than twenty were missing in all the action. But in the fort, not a single soldier escaped without a wound; and in one cohort, four centurions lost their eyes. And being desirous to produce testimony of the fatigue they underwent, and the danger they sustained, they counted to Caesar about thirty thousand arrows which had been thrown into the fort; and in the shield of the centurion Scaeva, which was brought to him, were found two hundred and thirty holes. In reward for this man's services both to himself and the republic, Caesar presented to him two hundred thousand pieces of copper money, and declared him promoted from the eighth to the first centurion. For it appeared that the fort had been in a great measure saved by his exertions; and he afterwards very amply rewarded the cohorts with double pay, corn, clothing, and other military honours.

LIV. — Pompey, having made great additions to his works in the night, the following days built turrets, and having carried his works fifteen feet high, faced that part of his camp with mantlets; and after an interval of five days, taking advantage of a second cloudy night, he barricaded all the gates of his camp to hinder a pursuit, and about midnight quietly marched off his army, and retreated to his old fortifications.

LV. — Aetolia, Acarnania, and Amphilochis, being reduced, as we have related, by Cassius Longinus, and Calvisius Sabinus, Caesar thought he ought to attempt the conquest of Achaia, and to advance farther into the country. Accordingly, he detached Fufius thither, and ordered Quintus Sabinus and Cassius to join him with their cohorts. Upon notice of their approach, Rutilius Lupus, who commanded in Achaia, under Pompey, began to fortify the Isthmus, to prevent Fufius from coming into Achaia. Kalenus recovered Delphi, Thebes, and Orchomenus, by a voluntary submission of those states. Some he subdued by force, the rest he endeavoured to win over to Caesar's interest, by sending deputies round to them. In these things, principally, Fufius was employed.

LVI. — Every day afterwards, Caesar drew up his army on a level ground, and offered Pompey battle, and led his legions almost close to Pompey's camp; and his front line was at no greater distance from the rampart than that no weapons from their engines could reach it. But Pompey, to save his credit and reputation with the world, drew out his legions, but so close to his camp that his rear lines might touch the rampart, and that his whole army, when drawn up, might be protected by the darts discharged from it.

LVII. — Whilst these things were going forward in Achaia and at Dyrrachium, and when it was certainly known that Scipio was arrived in Macedonia, Caesar, never losing sight of his first intention, sends Clodius to him, an intimate friend to both, whom Caesar, on the introduction and recommendation of Pompey, had admitted into the number of his acquaintance. To this man he gave letters and instructions to Pompey, the substance of which was as follows: "That he had made every effort towards peace, and imputed the ill success of those efforts to the fault of those whom he had employed to conduct those negotiations: because they were afraid to carry his proposals to Pompey at an improper time. That Scipio had such authority, that he could not only freely explain what conduct met his approbation, but even in some degree enforce his advice, and govern him [Pompey] if he persisted in error; that he commanded an army independent of Pompey, so that besides his authority, he had strength to compel; and if he did so, all men would be indebted to him for the quiet of Italy, the peace of the provinces, and the preservation of the empire." These proposals Clodius made to him, and for some days at the first appeared to have met with a favourable reception, but afterwards was not admitted to an audience; for Scipio being reprimanded by Favonius, as we found afterwards when the war was ended, and the negotiation having miscarried, Clodius returned to Caesar.

LVIII. — Caesar, that he might the more easily keep Pompey's horse enclosed within Dyrrachium, and prevent them from foraging, fortified the two narrow passes already mentioned with strong works, and erected forts at them. Pompey perceiving that he derived no advantage from his cavalry, after a few days had them conveyed back to his camp by sea. Fodder was so exceedingly scarce that he was obliged to feed his horses upon leaves stripped off the trees, or the tender roots of reeds pounded. For the corn which had been sown within the lines was already consumed, and they would be obliged to supply themselves with fodder from Corcyra and Acarnania, over a long tract of sea; and as the quantity of that fell short, to increase it by mixing barley with it, and by these methods support their cavalry. But when not only the barley and fodder in these parts were consumed, and the herbs cut away, when the leaves too were not to be found on the trees, the horses being almost starved, Pompey thought he ought to make some attempt by a sally.

LIX. — In the number of Caesar's cavalry were two Allobrogians, brothers, named Roscillus and Aegus, the sons of Abducillus, who for several years possessed the chief power in his own state; men of singular valour, whose gallant services Caesar had found very useful in all his wars in Gaul. To them, for these reasons, he had committed the offices of greatest honour in their own country, and took care to have them chosen into the senate at an unusual age, and had bestowed on them lands taken from the enemy, and large pecuniary rewards, and from being needy had made them affluent. Their valour had not only procured them Caesar's esteem, but they were beloved by the whole army. But presuming on Caesar's friendship, and elated with the arrogance natural to a foolish and barbarous people, they despised their countrymen, defrauded their cavalry of their pay, and applied all the plunder to their own use. Displeased at this conduct, their soldiers went in a body to Caesar, and openly complained of their ill usage; and to their other charges added, that false musters were given in to Caesar, and the surcharged pay applied to their own use.

LX. — Caesar, not thinking it a proper time to call them to account, and willing to pardon many faults, on account of their valour, deferred the whole matter, and gave them a private rebuke, for having made a traffic of their troops, and advised them to expect everything from his friendship, and by his past favours to measure their future hopes. This, however, gave them great offence, and made them contemptible in the eyes of the whole army. Of this they became sensible, as well from the reproaches of others, as from the judgment of their own minds, and a consciousness of guilt. Prompted then by shame, and perhaps imagining that they were not liberated from trial, but reserved to a future day, they resolved to break off from us, to put their fortune to a new hazard, and to make trial of new connections. And having conferred with a few of their clients, to whom they could venture to entrust so base an action, they first attempted to assassinate Caius Volusenus, general of the horse (as was discovered at the end of the war), that they might appear to have fled to Pompey after conferring an important service on him. But when that appeared too difficult to put in execution, and no opportunity offered to accomplish it, they borrowed all the money they could, as if they designed to make satisfaction and restitution for what they had defrauded: and having purchased a great number of horses, they deserted to Pompey along with those whom they had engaged in their plot.

LXI. — As they were persons nobly descended and of liberal education, and had come with a great retinue, and several cattle, and were reckoned men of courage, and had been in great esteem with Caesar, and as it was a new and uncommon event, Pompey carried them round all his works, and made an ostentatious show of them, for till that day, not a soldier, either horse or foot, had deserted from Caesar to Pompey, though there were desertions almost every day from Pompey to Caesar: but more commonly among the soldiers levied in Epirus and Aetolia, and in those countries which were in Caesar's possession. But the brothers, having been acquainted with all things, either what was incomplete in our works, or what appeared to the best judges of military matters to be deficient, the particular times, the distance of places, and the various attention of the guards, according to the different temper and character of the officer who commanded the different posts, gave an exact account of all to Pompey.

LXII. — Upon receiving this intelligence, Pompey, who had already formed the design of attempting a sally, as before mentioned, ordered the soldiers to make ozier coverings for their helmets, and to provide fascines. These things being prepared, he embarked on board small boats and row galleys by night, a considerable number of light infantry and archers, with all their fascines, and immediately after midnight, he marched sixty cohorts drafted from the greater camp and the outposts, to that part of our works which extended towards the sea, and were at the farthest distance from Caesar's greater camp. To the same place he sent the ships, which he had freighted with the fascines and light-armed troops; and all the ships of war that lay at Dyrrachium; and to each he gave particular instructions: at this part of the lines Caesar had posted Lentulus Marcellinus, the quaestor, with the ninth legion, and as he was not in a good state of health, Fulvius Costhumus was sent to assist him in the command.

LXIII. — At this place, fronting the enemy, there was a ditch fifteen feet wide, and a rampart ten feet high, and the top of the rampart was ten feet in breadth. At an interval of six hundred feet from that there was another rampart turned the contrary way, with the works lower. For some days before, Caesar, apprehending that our men might be surrounded by sea, had made a double rampart there, that if he should be attacked on both sides, he might have the means in defending himself. But the extent of the lines, and the incessant labour for so many days, because he had enclosed a circuit of seventeen miles with his works, did not allow time to finish them. Therefore the transverse rampart which should make a communication between the other two, was not yet completed. This circumstance was known to Pompey, being told to him by the Allobrogian deserters, and proved of great disadvantage to us. For when our cohorts of the ninth legion were on guard by the sea-side, Pompey's army arrived suddenly by break of day, and their approach was a surprise to our men, and at the same time, the soldiers that came by sea cast their darts on the front rampart; and the ditches were filled with fascines: and the legionary soldiers terrified those that defended the inner rampart, by applying the scaling ladders, and by engines and weapons of all sorts, and a vast multitude of archers poured round upon them from every side. Besides, the coverings of oziers, which they had laid over their helmets, were a great security to them against the blows of stones which were the only weapons that our soldiers had. And therefore, when our men were oppressed in every manner, and were scarcely able to make resistance, the defect in our works was observed, and Pompey's soldiers, landing between the two ramparts, where the work was unfinished, attacked our men in the rear, and having beat them from both sides of the fortification, obliged them to flee.

LXIV. — Marcellinus, being informed of this disorder, detached some cohorts to the relief of our men, who seeing them flee from the camp, were neither able to persuade them to rally at their approach, nor themselves to sustain the enemy's charge. And in like manner, whatever additional assistance was sent, was infected by the fears of the defeated, and increased the terror and danger. For retreat was prevented by the multitude of the fugitives. In that battle, when the eagle-bearer was dangerously wounded, and began to grow weak, having got sight of our horse, he said to them, "This eagle have I defended with the greatest care for many years, at the hazard of my life, and now in my last moments restore it to Caesar with the same fidelity. Do not, I conjure you, suffer a dishonour to be sustained in the field, which never before happened to Caesar's army, but deliver it safe into his hands." By this accident the eagle was preserved, but all the centurions of the first cohorts were killed, except the principal.

LXV. — And now the Pompeians, after great havoc of our troops, were approaching Marcellinus's camp, and had struck no small terror into the rest of the cohorts, when Marcus Antonius, who commanded the nearest fort, being informed of what had happened, was observed descending from the rising ground with twelve cohorts. His arrival checked the Pompeians, and encouraged our men to recover from their extreme affright. And shortly after, Caesar having got notice by the smoke from all the forts, which was the usual signal on such occasions, drafted off some cohorts from the outposts, and went to the scene of action. And having there learnt the loss he had sustained, and perceiving that Pompey had forced our works, and had encamped along the coast, so that he was at liberty to forage, and had a communication with his shipping, he altered his plan for conducting the war, as his design had not succeeded, and ordered a strong encampment to be made near Pompey.

LXVI. — When this work was finished, Caesar's scouts observed that some cohorts, which to them appeared like a legion, were retired behind the wood, and were on their march to the old camp. The situation of the two camps was as follows: a few days before, when Caesar's ninth legion had opposed a party of Pompey's troops, and were endeavouring to enclose them, Caesar's troops formed a camp in that place. This camp joined a certain wood, and was not above four hundred paces distant from the sea. Afterwards, changing his design for certain reasons, Caesar removed his camp to a small distance beyond that place; and after a few days, Pompey took possession of it, and added more extensive works, leaving the inner rampart standing, as he intended to keep several legions there. By this means, the lesser camp included within the greater, answered the purpose of a fort and citadel. He had also carried an entrenchment from the left angle of the camp to the river, about four hundred paces, that his soldiers might have more liberty and less danger in fetching water. But he too, changing his design for reasons not necessary to be mentioned, abandoned the place. In this condition the camp remained for several days, the works being all entire.

LXVII. — Caesar's scouts brought him word that the standard of a legion was carried to this place. That the same thing was seen he was assured by those in the higher forts. This place was half a mile distant from Pompey's new camp. Caesar, hoping to surprise this legion, and anxious to repair the loss sustained that day, left two cohorts employed in the works to make an appearance of entrenching himself, and by a different route, as privately as he could, with his other cohorts amounting to thirty-three, among which was the ninth legion, which had lost so many centurions, and whose privates were greatly reduced in number, he marched in two lines against Pompey's legion and his lesser camp. Nor did this first opinion deceive him. For he reached the place before Pompey could have notice of it; and though the works were strong, yet having made the attack with the left wing, which he commanded in person, he obliged the Pompeians to quit the rampart in disorder. A barricade had been raised before the gates, at which a short contest was maintained, our men endeavouring to force their way in, and the enemy to defend the camp; Titus Pulcio, by whose means we have related that Caius Antonius's army was betrayed, defending them with singular courage. But the valour of our men prevailed, and having cut down the barricade, they first forced the greater camp, and after that the fort which was enclosed within it: and as the legion on its repulse had retired to this, they slew several defending themselves there.

LXVIII. — But Fortune, who exerts a powerful influence as well in other matters, as especially in war, effects great changes from trifling causes, as happened at this time. For the cohorts on Caesar's right wing, through ignorance of the place, followed the direction of that rampart, which ran along from the camp to the river, whilst they were in search of a gate, and imagined that it belonged to the camp. But when they found that it led to the river, and that nobody opposed them, they immediately climbed over the rampart, and were followed by all our cavalry.

LXIX. — In the meantime Pompey, by the great delay which this occasioned, being informed of what had happened, marched with the fifth legion, which he called away from their work to support his party; and at the same time his cavalry were advancing up to ours, and an army in order of battle was seen at a distance by our men who had taken possession of the camp, and the face of affairs was suddenly changed. For Pompey's legion, encouraged by the hope of speedy support, attempted to make a stand at the Decuman gate, and made a bold charge on our men. Caesar's cavalry, who had mounted the rampart by a narrow breach, being apprehensive of their retreat, were the first to flee. The right wing, which had been separated from the left, observing the terror of the cavalry, to prevent their being overpowered within the lines, were endeavouring to retreat by the same way as they burst in; and most of them, lest they should be engaged in the narrow passes, threw themselves down a rampart ten feet high into the trenches; and the first being trodden to death, the rest procured their safety and escaped over their bodies. The soldiers of the left wing, perceiving from the rampart that Pompey was advancing, and their own friends fleeing, being afraid that they should be enclosed between the two ramparts, as they had an enemy both within and without, strove to secure their retreat the same way they came. All was disorder, consternation, and flight; insomuch that, when Caesar laid hold of the colours of those who were running away, and desired them to stand, some left their horses behind, and continued to run in the same manner; others through fear even threw away their colours, nor did a single man face about.

LXX. — In this calamity, the following favourable circumstance occurred to prevent the ruin of our whole army, viz., that Pompey suspecting an ambuscade (because, as I suppose, the success had far exceeded his hopes, as he had seen his men a moment before fleeing from the camp), durst not for some time approach the fortification; and that his horse were retarded from pursuing, because the passes and gates were in possession of Caesar's soldiers. Thus a trifling circumstance proved of great importance to each party; for the rampart drawn from the camp to the river, interrupted the progress and certainty of Caesar's victory, after he had forced Pompey's camp. The same thing, by retarding the rapidity of the enemy's pursuit, preserved our army.

LXXI. — In the two actions of this day, Caesar lost nine hundred and sixty rank and file, several Roman knights of distinction, Felginas Tuticanus Gallus, a senator's son; Caius Felginas from Placentia; Aulus Gravius from Puteoli; Marcus Sacrativir from Capua; and thirty-two military tribunes and centurions. But the greatest part of all these perished without a wound, being trodden to death in the trenches, on the ramparts and banks of the river by reason of the terror and flight of their own men. Pompey, after this battle, was saluted Imperator; this title he retained, and allowed himself to be addressed by it afterwards. But neither in his letters to the senate, nor in the fasces, did he use the laurel as a mark of honour. But Labienus, having obtained his consent that the prisoners should be delivered up to him, had them all brought out, as it appeared, to make a show of them, and that Pompey might place a greater confidence in him who was a deserter; and calling them fellow soldiers, and asking them in the most insulting manner whether it was usual with veterans to flee, ordered them to be put to death in the sight of the whole army.

LXXII. — Pompey's party were so elated with confidence and spirit at this success, that they thought no more of the method of conducting the war, but thought that they were already conquerors. They did not consider that the smallness of our numbers, and the disadvantage of the place and the confined nature of the ground occasioned by their having first possessed themselves of the camp, and the double danger both from within and without the fortifications, and the separation of the army into two parts, so that the one could not give relief to the other, were the cause of our defeat. They did not consider, in addition, that the contest was not decided by a vigorous attack, nor a regular battle; and that our men had suffered greater loss from their numbers and want of room, than they had sustained from the enemy. In fine, they did not reflect on the common casualties of war; how trifling causes, either from groundless suspicions, sudden affright, or religious scruples, have oftentimes been productive of considerable losses; how often an army has been unsuccessful either by the misconduct of the general, or the oversight of a tribune; but as if they had proved victorious by their valour, and as if no change could ever take place, they published the success of the day throughout the world by reports and letters.

LXXIII. — Caesar, disappointed in his first intentions, resolved to change the whole plan of his operations. Accordingly, he at once called in all out-posts, gave over the siege, and collecting his army into one place, addressed his soldiers and encouraged them "not to be troubled at what had happened, nor to be dismayed at it, but to weigh their many successful engagements against one disappointment, and that, too, a trifling one.

That they ought to be grateful to Fortune, through whose favour they had recovered Italy without the effusion of blood; through whose favour they had subdued the two Spains, though protected by a most warlike people under the command of the most skilful and experienced generals: through whose favour they had reduced to submission the neighbouring states that abounded with corn: in fine, that they ought to remember with what success they had been all transported safe through blockading fleets of the enemy, which possessed not only the ports, but even the coasts: that if all their attempts were not crowned with success, the defects of Fortune must be supplied by industry; and whatever loss had been sustained, ought to be attributed rather to her caprices than to any faults in him: that he had chosen a safe ground for the engagement, that he had possessed himself of the enemy's camp; that he had beaten them out, and overcome them when they offered resistance; but whether their own terror or some mistake, or whether Fortune herself had interrupted a victory almost secured and certain, they ought all now to use their utmost efforts to repair by their valour the loss which had been incurred; if they did so, their misfortunes would turn to their advantage, as it happened at Gergovia, and those who feared to face the enemy would be the first to offer themselves to battle.

LXXIV.— Having concluded his speech, he disgraced some standard-bearers, and reduced them to the ranks; for the whole army was seized with such grief at their loss, and with such an ardent desire of repairing their disgrace, that not a man required the command of his tribune or centurion, but they imposed each on himself severer labours than usual as a punishment, and at the same time were so inflamed with eagerness to meet the enemy, that the officers of the first rank, sensibly affected at their entreaties, were of opinion that they ought to continue in their present posts, and commit their fate to the hazard of a battle. But, on the other hand, Caesar could not place sufficient confidence in men so lately thrown into consternation, and thought he ought to allow them time to recover their dejected spirits; and having abandoned his works, he was apprehensive of being distressed for want of corn.

LXXV.— Accordingly, suffering no time to intervene but what was necessary for a proper attention to be paid to the sick and wounded, he sent on all his baggage privately in the beginning of the night from his camp to Apollonia, and ordered them not to halt till they had performed their journey; and he detached one legion with them as a convoy. This affair being concluded, having retained only two legions in his camp; he marched the rest of his army out at three o'clock in the morning by several gates, and sent them forward by the same route; and in a short space after, that the military practice might be preserved, and his march known as late as possible, he ordered the signal for decamping to be given; and setting out immediately, and following the rear of his own army, he was soon out of sight of the camp. Nor did Pompey, as soon as he had notice of his design, make any delay to pursue him; but with a view to surprise them whilst encumbered with baggage on their march, and not yet recovered from their fright, he led his army out of his camp, and sent his cavalry on to retard our rear; but was not able to come up with them, because Caesar had got far before him, and marched without baggage. But when we reached the river Genusus, the banks being steep, their horse overtook our rear, and detained them by bringing them to action. To oppose whom, Caesar sent his horse, and intermixed with them about four hundred of his advanced light troops, who attacked their horse with such success, that having routed them all, and killed several, they returned without any loss to the main body.

LXXVI. — Having performed the exact march which he had proposed that day, and having led his army over the river Genusus, Caesar posted himself in his old camp opposite Asparagium; and kept his soldiers close within the entrenchments; and ordered the horse, who had been sent out under pretence of foraging, to retire immediately into the camp, through the Decuman gate. Pompey, in like manner, having completed the same day's march, took post in his old camp at Asparagium; and his soldiers, as they had no work (the fortifications being entire), made long excursions, some to collect wood and forage; others, invited by the nearness of the former camp, laid up their arms in their tents, and quitted the entrenchments in order to bring what they had left behind them, because the design of marching being adopted in a hurry, they had left a considerable part of their waggons and luggage behind. Being thus incapable of pursuing, as Caesar had foreseen, about noon he gave the signal for marching, led out his army, and doubling that day's march, he advanced eight miles beyond Pompey's camp; who could not pursue him, because his troops were dispersed.

LXXVII. — The next day Caesar sent his baggage forward early in the night, and marched off himself immediately after the fourth watch: that if he should be under the necessity of risking an engagement, he might meet a sudden attack with an army free from incumbrance. He did so for several days successively, by which means he was enabled to effect his march over the deepest rivers, and through the most intricate roads without any loss. For Pompey, after the first day's delay, and the fatigue which he endured for some days in vain, though he exerted himself by forced marches, and was anxious to overtake us, who had got the start of him, on the fourth day desisted from the pursuit, and determined to follow other measures.

LXXVIII. — Caesar was obliged to go to Apollonia, to lodge his wounded, pay his army, confirm his friends, and leave garrisons in the towns. But for these matters, he allowed no more time than was necessary for a person in haste. And being apprehensive for Domitius, lest he should be surprised by Pompey's arrival, he hastened with all speed and earnestness to join him; for he planned the operations of the whole campaign on these principles: that if Pompey should march after him, he would be drawn off from the sea, and from those forces which he had provided in Dyrrachium, and separated from his corn and magazines, and be obliged to carry on the war on equal terms; but if he crossed over into Italy, Caesar, having effected a junction with Domitius, would march through Illyricum to the relief of Italy; but if he endeavoured to storm Apollonia and Oricum, and exclude him from the whole coast, he hoped, by besieging Scipio, to oblige him, of necessity, to come to his assistance. Accordingly, Caesar despatching couriers, writes to Domitius, and acquaints him with his wishes on the subject: and having stationed a garrison of four cohorts at Apollonia, one at Lissus, and three at Oricum, besides those who were sick of their wounds, he set forward on his march through Epirus and Acarnania. Pompey, also, guessing at Caesar's design, determined to hasten to Scipio, that if Caesar should march in that direction, he might be ready to relieve him; but that if Caesar should be unwilling to quit the sea-coast and Corcyra, because he expected legions and cavalry from Italy, he himself might fall on Domitius with all his forces.

LXXIX. — For these reasons, each of them studied despatch, that he might succour his friends, and not miss an opportunity of surprising his enemies. But Caesar's engagements at Apolloma had carried him aside from the direct road. Pompey had taken the short road to Macedonia, through Candavia. To this was added another unexpected disadvantage, that Domitius, who for several days had been encamped opposite Scipio, had quitted that post for the sake of provisions, and had marched to Heraclea Sentica, a city subject to Candavia; so that fortune herself seemed to throw him in Pompey's way. Of this, Caesar was ignorant up to this time. Letters likewise being sent by Pompey through all the provinces and states, with an account of the action at Dyrrachium, very much enlarged and exaggerated beyond the real facts, a rumour had been circulated, that Caesar had been defeated and forced to flee, and had lost almost all his forces. These reports had made the roads dangerous, and drawn off some states from his alliance: whence it happened, that the messengers despatched by Caesar, by several different roads to Domitius, and by Domitius to Caesar, were not able by any means to accomplish their journey. But the Allobroges, who were in the retinue of Aegus and Roscillus, and who had deserted to Pompey, having met on the road a scouting party of Domitius; either from old acquaintance, because they had served together in Gaul, or elated with vainglory, gave them an account of all that had happened, and informed them of Caesar's departure, and Pompey's arrival. Domitius, who was scarce four hours' march distant, having got intelligence from these, by the courtesy of the enemy, avoided the danger, and met Caesar coming to join him at Aeginium, a town on the confines of and opposite to Thessaly.

LXXX. — The two armies being united, Caesar marched to Gomphi, which is the first town of Thessaly on the road from Epirus. Now, the Thessalians, a few months before, had of themselves sent ambassadors to Caesar, offering him the free use of everything in their power, and requesting a garrison for their protection. But the report, already spoken of, of the battle at Dyrrachium, which it had exaggerated in many particulars, had arrived before him. In consequence of which, Androsthenes, the praetor of Thessaly, as he preferred to be the companion of Pompey's victory, rather than Caesar's associate in his misfortunes, collected all the people, both slaves and freemen, from the country into the town and shut the gates, and despatched messengers to Scipio and Pompey "to come to his relief, that he could depend on the strength of the town, if succour was speedily sent; but that it could not withstand a long siege." Scipio, as soon as he received advice of the departure of the armies from Dyrrachium, had marched with his legions to Larissa: Pompey was not yet arrived near Thessaly. Caesar having fortified his camp, ordered scaling ladders and penthouses to be made for a sudden assault, and hurdles to be provided. As soon as they were ready, he exhorted his soldiers, and told them of what advantage it would be to assist them with all sorts of necessaries if they made themselves masters of a rich and plentiful town: and, at the same time, to strike terror into other states by the example of this, and to effect this with speed, before auxiliaries could arrive. Accordingly, taking advantage of the unusual ardour of the soldiers, he began his assault on the town at a little after three o'clock on the very day on which he arrived, and took it, though defended with very high walls, before sunset, and gave it up to his army to plunder, and immediately decamped from before it, and marched to Metropolis, with such rapidity as to outstrip any messenger or rumour of the taking of Gomphi.

LXXXI. — The inhabitants of Metropolis, at first influenced by the same rumours, followed the same measures, shut the gates and manned their walls. But when they were made acquainted with the fate of the city of Gomphi by some prisoners, whom Caesar had ordered to be brought up to the walls, they threw open their gates. As he preserved them with the greatest care, there was not a state in Thessaly (except Larissa, which was awed by a strong army of Scipio's), but on comparing the fate of the inhabitants of Metropolis with the severe treatment of Gomphi, gave admission to Caesar, and obeyed his orders. Having chosen a position convenient for procuring corn, which was now almost ripe on the ground, he determined there to wait Pompey's arrival, and to make it the centre of all his warlike operations.

LXXXII. — Pompey arrived in Thessaly a few days after, and having harangued the combined army, returned thanks to his own men, and exhorted Scipio's soldiers, that as the victory was now secured, they should endeavour to merit a part of the rewards and booty. And receiving all the legions into one camp, he shared his honours with Scipio, ordered the trumpet to be sounded at his tent, and a pavilion to be erected for him. The forces of Pompey being thus augmented, and two such powerful armies united, their former expectations were confirmed, and their hopes of victory so much increased, that whatever time intervened was considered as so much delay to their return into Italy: and whenever Pompey acted with slowness and caution, they used to exclaim, that it was the business only of a single day, but that he had a passion for power, and was delighted in having persons of consular and praetorian rank in the number of his slaves. And they now began to dispute openly about rewards and priesthoods, and disposed of the consulate for several years to come. Others put in their claims for the houses and properties of all who were in Caesar's camp, and in that council there was a warm debate, whether Lucius Hirrus, who had been sent by Pompey against the Parthians, should be admitted a candidate for the praetorship in his absence at the next election; his friends imploring Pompey's honour to fulfil the engagements which he had made to him at his departure, that he might not seem deceived through his authority: whilst others, embarked in equal labour and danger, pleaded that no individual ought to have a preference before all the rest.

LXXXIII. — Already Domitius, Scipio, and Lentulus Spinthur, in their daily quarrels about Caesar's priesthood, openly abused each other in the most scurrilous language. Lentulus urging the respect due to his age, Domitius boasting his interest in the city and his dignity, and Scipio presuming on his alliance with Pompey. Attius Rufus charged Lucius Afranius before Pompey with betraying the army in the action that happened in Spain, and Lucius Domitius declared in the council that it was his wish that, when the war should be ended, three billets should be given to all the senators who had taken part with them in the war, and that they should pass sentence on every single person who had stayed behind at Rome, or who had been within Pompey's garrisons and had not contributed their assistance in the military operations; that by the first billet they should-have power to acquit, by the second to pass sentence of death, and by the third to impose a pecuniary fine. In short, Pompey's whole army talked of nothing but the honours or sums of money which were to be their rewards, or of vengeance on their enemies; and never considered how they were to defeat their enemies, but in what manner they should use their victory.

LXXXIV. — Corn being provided, and his soldiers refreshed, and a sufficient time having elapsed since the engagement at Dyrrachium, when Caesar thought he had sufficiently sounded the disposition of his troops, he thought that he ought to try whether Pompey had any intention or inclination to come to a battle. Accordingly he led his troops out of the camp, and ranged them in order of battle, at first on their own ground, and at a small distance from Pompey's camp: but afterwards for several days in succession he advanced from his own camp, and led them up to the hills on which Pompey's troops were posted, which conduct inspired his army every day with fresh courage. However he adhered to his former purpose respecting his cavalry, for as he was by many degrees inferior in number, he selected the youngest and most active of the advanced guard, and desired them to fight intermixed with the horse, and they by constant practice acquired experience in this kind of battle. By these means it was brought to pass that a thousand of his horse would dare, even on open ground, to stand against seven thousand of Pompey's, if occasion required, and would not be much terrified by their number. For even on one of those days he was successful in a cavalry action, and killed one of the two Allobrogians who had deserted to Pompey, as we before observed, and several others.

LXXXV. — Pompey, because he was encamped on a hill, drew up his army at the very foot of it, ever in expectation, as may be conjectured, that Caesar would expose himself to this disadvantageous situation. Caesar, seeing no likelihood of being able to bring Pompey to an action, judged it the most expedient method of conducting the war, to decamp from that post, and to be always in motion: with this hope, that by shifting his camp and removing from place to place, he might be more conveniently supplied with corn, and also, that by being in motion he might get some opportunity of forcing them to battle, and might by constant marches harass Pompey's army, which was not accustomed to fatigue. These matters being settled, when the signal for marching was given, and the tents struck, it was observed that shortly before, contrary to his daily practice, Pompey's army had advanced farther than usual from his entrenchments, so that it appeared possible to come to an action on equal ground. Then Caesar addressed himself to his soldiers, when they were at the gates of the camp, ready to march out. "We must defer," says he, "our march at present, and set our thoughts on battle, which has been our constant wish; let us then meet the foe with resolute souls. We shall not hereafter easily find such an opportunity." He immediately marched out at the head of his troops.

LXXXVI. — Pompey also, as was afterward known, at the unanimous solicitation of his friends, had determined to try the fate of a battle. For he had even declared in council a few days before that, before the battalions came to battle, Caesar's army would be put to the rout. When most people expressed their surprise at it, "I know," says he, "that I promise a thing almost incredible; but hear the plan on which I proceed, that you may march to battle with more confidence and resolution. I have persuaded our cavalry, and they have engaged to execute it, as soon as the two armies have met, to attack Caesar's right wing on the flank, and enclosing their army on the rear, throw them into disorder, and put them to the rout, before we shall throw a weapon against the enemy. By this means we shall put an end to the war, without endangering the legions, and almost without a blow. Nor is this a difficult matter, as we far outnumber them in cavalry." At the same time he gave them notice to be ready for battle on the day following, and since the opportunity which they had so often wished for was now arrived, not to disappoint the opinion generally entertained of their experience and valour.

LXXXVII. — After him Labienus spoke, as well to express his contempt of Caesar's forces, as to extol Pompey's scheme with the highest encomiums. "Think not, Pompey," says he, "that this is the army which conquered Gaul and Germany; I was present at all those battles and do not speak at random on a subject to which I am a stranger: a very small part of that army now remains, great numbers lost their lives, as must necessarily happen in so many battles, many fell victims to the autumnal pestilence in Italy, many returned home, and many were left behind on the continent. Have you not heard that the cohorts at Brundisium are composed of invalids? The forces which you now behold, have been recruited by levies lately made in Hither Spain, and the greater part from the colonies beyond the Po; moreover, the flower of the forces perished in the two engagements at Dyrrachium." Having so said, he took an oath, never to return to his camp unless victorious; and he encouraged the rest to do the like. Pompey applauded his proposal, and took the same oath; nor did any person present hesitate to take it. After this had passed in the council they broke up full of hopes and joy, and in imagination anticipated victory; because they thought that in a matter of such importance, no groundless assertion could be made by a general of such experience.

LXXXVIII. — When Caesar had approached near Pompey's camp, he observed that his army was drawn up in the following manner: — On the left wing were the two legions delivered over by Caesar at the beginning of the disputes in compliance with the senate's decree, one of which was called the first, the other the third. Here Pompey commanded in person. Scipio with the Syrian legions commanded the centre. The Cilician legion in conjunction with the Spanish cohorts, which we said were brought over by Afranius, were disposed on the right wing. These Pompey considered his steadiest troops. The rest he had interspersed between the centre and the wing, and he had a hundred and ten complete cohorts; these amounted to forty-five thousand men. He had besides two cohorts of volunteers, who having received favours from him in former wars, flocked to his standard: these were dispersed through his whole army. The seven remaining cohorts he had disposed to protect his camp, and the neighbouring forts. His right wing was secured by a river with steep banks; for which reason he placed all his cavalry, archers, and slingers, on his left wing.

LXXXIX. — Caesar, observing his former custom, had placed the tenth legion on the right, the ninth on the left, although it was very much weakened by the battles at Dyrrachium. He placed the eighth legion so close to the ninth, as to almost make one of the two, and ordered them to support one another. He drew up on the field eighty cohorts, making a total of twenty-two thousand men. He left two cohorts to guard the camp. He gave the command of the left wing to Antonius, of the right to P. Sulla, and of the centre to Cn. Domitius: he himself took his post opposite Pompey. At the same time, fearing, from the disposition of the enemy which we have previously mentioned, lest his right wing might be surrounded by their numerous cavalry, he rapidly drafted a single cohort from each of the legions composing the third line, formed of them a fourth line, and opposed them to Pompey's cavalry, and, acquainting them with his wishes, admonished them that the success of that day depended on their courage. At the same time he ordered the third line, and the entire army not to charge without his command: that he would give the signal whenever he wished them to do so.

XC. — When he was exhorting his army to battle, according to the military custom, and spoke to them of the favours that they had constantly received from him, he took especial care to remind them "that he could call his soldiers to witness the earnestness with which he had sought peace, the efforts that he had made by Vatinius to gain a conference [with Labienus], and likewise by Claudius to treat with Scipio, in what manner he had exerted himself at Oricum, to gain permission from Libo to send ambassadors; that he had been always reluctant to shed the blood of his soldiers, and did not wish to deprive the republic of one or other of her armies." After delivering this speech, he gave by a trumpet the signal to his soldiers, who were eagerly demanding it, and were very impatient for the onset.

XCI. — There was in Caesar's army a volunteer of the name of Crastinus, who the year before had been first centurion of the tenth legion, a man of pre-eminent bravery. He, when the signal was given, says, "Follow me, my old comrades, and display such exertions in behalf of your general as you have determined to do: this is our last battle, and when it shall be won, he will recover his dignity, and we our liberty." At the same time he looked back to Caesar, and said, "General, I will act in such a manner to-day, that you will feel grateful to me living or dead." After uttering these words he charged first on the right wing, and about one hundred and twenty chosen volunteers of the same century followed.

XCII. — There was so much space left between the two lines, as sufficed for the onset of the hostile armies: but Pompey had ordered his soldiers to await Caesar's attack, and not to advance from their position, or suffer their line to be put into disorder. And he is said to have done this by the advice of Caius Triarius, that the impetuosity of the charge of Caesar's soldiers might be checked, and their line broken, and that Pompey's troops remaining in their ranks, might attack them while in disorder; and he thought that the javelins would fall with less force if the soldiers were kept in their ground, than if they met them in their course; at the same time he trusted that Caesar's soldiers, after running over double the usual ground, would become weary and exhausted by the fatigue. But to me Pompey seems to have acted without sufficient reason: for there is a certain impetuosity of spirit and an alacrity implanted by nature in the hearts of all men, which is inflamed by a desire to meet the foe. This a general should endeavour not to repress, but to increase; nor was it a vain institution of our ancestors, that the trumpets should sound on all sides, and a general shout be raised; by which they imagined that the enemy were struck with terror, and their own army inspired with courage.

XCIII. — But our men, when the signal was given, rushed forward with their javelins ready to be launched, but perceiving that Pompey's men did not run to meet their charge, having acquired experience by custom, and being practised in former battles, they of their own accord repressed their speed, and halted almost midway, that they might not come up with the enemy when their strength was exhausted, and after a short respite they again renewed their course, and threw their javelins, and instantly drew their swords, as Caesar had ordered them. Nor did Pompey's men fail in this crisis, for they received our javelins, stood our charge, and maintained their ranks: and having launched their javelins, had recourse to their swords. At the same time Pompey's horse, according to their orders, rushed out at once from his left wing, and his whole host of archers poured after them. Our cavalry did not withstand their charge: but gave ground a little, upon which Pompey's horse pressed them more vigorously, and began to file off in troops, and flank our army. When Caesar perceived this, he gave the signal to his fourth line, which he had formed of the six cohorts. They instantly rushed forward and charged Pompey's horse with such fury, that not a man of them stood; but all wheeling about, not only quitted their post, but galloped forward to seek a refuge in the highest mountains. By their retreat the archers and slingers, being left destitute and defenceless, were all cut to pieces. The cohorts, pursuing their success, wheeled about upon Pompey's left wing, whilst his infantry still continued to make battle, and attacked them in the rear.

XCIV. — At the same time Caesar ordered his third line to advance, which till then had not been engaged, but had kept their post. Thus, new and fresh troops having come to the assistance of the fatigued, and others having made an attack on their rear, Pompey's men were not able to maintain their ground, but all fled, nor was Caesar deceived in his opinion that the victory, as he had declared in his speech to his soldiers, must have its beginning from those six cohorts which he had placed as a fourth line to oppose the horse. For by them the cavalry were routed; by them the archers and slingers were cut to pieces; by them the left wing of Pompey's army was surrounded, and obliged to be the first to flee. But when Pompey saw his cavalry routed, and that part of his army on which he reposed his greatest hopes thrown into confusion, despairing of the rest, he quitted the field, and retreated straightway on horseback to his camp, and calling to the centurions, whom he had placed to guard the praetorian gate, with a loud voice, that the soldiers might hear: "Secure the camp," says he, "defend it with diligence, if any danger should threaten it; I will visit the other gates, and encourage the guards of the camp." Having thus said, he retired into his tent in utter despair, yet anxiously waiting the issue.

XCV. — Caesar having forced the Pompeians to flee into their entrenchment, and thinking that he ought not to allow them any respite to recover from their fright, exhorted his soldiers to take advantage of fortune's kindness, and to attack the camp. Though they were fatigued by the intense heat, for the battle had continued till mid-day, yet, being prepared to undergo any labour, they cheerfully obeyed his command. The camp was bravely defended by the cohorts which had been left to guard it, but with much more spirit by the Thracians and foreign auxiliaries. For the soldiers who had fled for refuge to it from the field of battle, affrighted and exhausted by fatigue, having thrown away their arms and military standards, had their thoughts more engaged on their further escape than on the defence of the camp. Nor could the troops who were posted on the battlements long withstand the immense number of our darts, but fainting under their wounds, quitted the place, and under the conduct of their centurions and tribunes, fled, without stopping, to the high mountains which joined the camp.

XCVI. — In Pompey's camp you might see arbours in which tables were laid, a large quantity of plate set out, the floors of the tents covered with fresh sods, the tents of Lucius Lentulus and others shaded with ivy, and many other things which were proofs of excessive luxury, and a confidence of victory, so that it might readily be inferred that they had no apprehensions of the issue of the day, as they indulged themselves in unnecessary pleasures, and yet upbraided with luxury Caesar's army, distressed and suffering troops, who had always been in want of common necessaries. Pompey, as soon as our men had forced the trenches, mounting his horse, and stripping off his general's habit, went hastily out of the back gate of the camp, and galloped with all speed to Larissa. Nor did he stop there, but with the same despatch collecting a few of his flying troops, and halting neither day nor night, he arrived at the sea-side, attended by only thirty horse, and went on board a victualling barque, often complaining, as we have been told, that he had been so deceived in his expectation, that he was almost persuaded that he had been betrayed by those from whom he had expected victory, as they began the flight.

XCVII. — Caesar having possessed himself of Pompey's camp, urged his soldiers not to be too intent on plunder, and lose the opportunity of completing their conquest. Having obtained their consent, he began to draw lines round the mountain. The Pompeians distrusting the position, as there was no water on the mountain, abandoned it, and all began to retreat towards Larissa; which Caesar perceiving, divided his troops, and ordering part of his legions to remain in Pompey's camp, sent back a part to his own camp, and taking four legions with him, went by a shorter road to intercept the enemy: and having marched six miles, drew up his army. But the Pompeians observing this, took post on a mountain whose foot was washed by a river. Caesar having encouraged his troops, though they were greatly exhausted by incessant labour the whole day, and night was now approaching, by throwing up works cut off the communication between the river and the mountain, that the enemy might not get water in the night. As soon as the work was finished, they sent ambassadors to treat about a capitulation. A few senators who had espoused that party, made their escape by night.

XCVIII. — At break of day, Caesar ordered all those who had taken post on the mountain, to come down from the higher grounds into the plain, and pile their arms. When they did this without refusal, and with outstretched arms, prostrating themselves on the ground, with tears, implored his mercy: he comforted them and bade them rise, and having spoken a few words of his own clemency to alleviate their fears, he pardoned them all, and gave orders to his soldiers that no injury should be done to them, and nothing taken from them. Having used this diligence, he ordered the legions in his camp to come and meet him, and those which were, with him to take their turn of rest, and go back to the camp; and the same day went to Larissa.

XCIX. — In that battle, no more than two hundred privates were missing, but Caesar lost about thirty centurions, valiant officers. Crastinus, also, of whom mention was made before, fighting most courageously, lost his life by the wound of a sword in the mouth; nor was that false which he declared when marching to battle: for Caesar entertained the highest opinion of his behaviour in that battle, and thought him highly deserving of his approbation. Of Pompey's army, there fell about fifteen thousand; but upwards of twenty-four thousand were made prisoners: for even the cohorts which were stationed in the forts, surrendered to Sylla. Several others took shelter in the neighbouring states. One hundred and eighty stands of colours, and nine eagles, were brought to Caesar. Lucius Domitius, fleeing from the camp to the mountains, his strength being exhausted by fatigue, was killed by the horse.

C. — About this time, Decimus Laelius arrived with his fleet at Brundisium and in the same manner as Libo had done before, possessed himself of an island opposite the harbour of Brundisium. In like manner, Valimus, who was then governor of Brundisium, with a few decked barques, endeavoured to entice Laelius's fleet, and took one five-benched galley and two smaller vessels that had ventured farther than the rest into a narrow part of the harbour: and likewise disposing the horse along the shore, strove to prevent the enemy from procuring fresh water. But Laelius having chosen a more convenient season of the year for his expedition, supplied himself with water brought in transports from Corcyra and Dyrrachium, and was not deterred from his purpose; and till he had received advice of the battle in Thessaly, he could not be forced either by the disgrace of losing his ships, or by the want of necessaries, to quit the port and islands.

CI. — Much about the same time, Cassius arrived in Sicily with a fleet of Syrians, Phoenicians, and Cilicians: and as Caesar's fleet was divided into two parts, Publius Sulpicius the praetor commanding one division at Vibo near the straits, Pomponius the other at Messana, Cassius got into Messana with his fleet before Pomponius had notice of his arrival, and having found him in disorder, without guards or discipline, and the wind being high and favourable, he filled several transports with fir, pitch, and tow, and other combustibles, and sent them against Pomponius's fleet, and set fire to all his ships, thirty-five in number, twenty of which were armed with beaks: and this action struck such terror, that though there was a legion in garrison at Messana, the town with difficulty held out, and had not the news of Caesar's victory been brought at that instant by the horse stationed along the coast, it was generally imagined that it would have been lost, but the town was maintained till the news arrived very opportunely; and Cassius set sail from thence to attack Sulpicius's fleet at Vibo, and our ships being moored to the land, to strike the same terror, he acted in the same manner as before. The wind being favourable, he sent into the port about forty ships provided with combustibles, and the flame catching on both sides, five ships were burnt to ashes. And when the fire began to spread wider by the violence of the wind, the soldiers of the veteran legions, who had been left to guard the fleet, being considered as invalids, could not endure the disgrace, but of themselves went on board the ships and weighed anchor, and having attacked Cassius's fleet, captured two five-banked galleys, in one of which was Cassius himself; but he made his escape by taking to a boat. Two three-banked galleys were taken besides. Intelligence was shortly after received of the action in Thessaly, so well authenticated, that the Pompeians themselves gave credit to it; for they had hitherto believed it a fiction of Caesar's lieutenants and friends. Upon which intelligence Cassius departed with his fleet from that coast.

CII. — Caesar thought he ought to postpone all business and pursue Pompey, whithersoever he should retreat; that he might not be able to provide fresh forces, and renew the war; he therefore marched on every day, as far as his cavalry were able to advance, and ordered one legion to follow him by shorter journeys. A proclamation was issued by Pompey at Amphipolis, that all the young men of that province, Grecians and Roman citizens, should take the military oath; but whether he issued it with an intention of preventing suspicion, and to conceal as long as possible his design of fleeing farther, or to endeavour to keep possession of Macedonia by new levies, if nobody pursued him, it is impossible to judge. He lay at anchor one night, and calling together his friends in Amphipolis, and collecting a sum of money for his necessary expenses, upon advice of Caesar's approach, set sail from that place, and arrived in a few days at Mitylene. Here he was detained two days, and having added a few galleys to his fleet he went to Cilicia, and thence to Cyprus. There he is informed that, by the consent of all the inhabitants of Antioch and Roman citizens who traded there, the castle had been seized to shut him out of the town; and that messengers had been despatched to all those who were reported to have taken refuge in the neighbouring states, that they should not come to Antioch; that if they did that, it would be attended with imminent danger to their lives. The same thing had happened to Lucius Lentulus, who had been consul the year before, and to Publius Lentulus a consular senator, and to several others at Rhodes, who having followed Pompey in his flight, and arrived at the island, were not admitted into the town or port; and having received a message to leave that neighbourhood, set sail much against their will; for the rumour of Caesar's approach had now reached those states.

CIII. — Pompey, being informed of these proceedings, laid aside his design of going to Syria, and having taken the public money from the farmers of the revenue, and borrowed more from some private friends, and having put on board his ships a large quantity of brass for military purposes, and two thousand armed men, whom he partly selected from the slaves of the tax farmers, and partly collected from the merchants, and such persons as each of his friends thought fit on this occasion, he sailed for Pelusium. It happened that king Ptolemy, a minor, was there with a considerable army, engaged in war with his sister Cleopatra, whom a few months before, by the assistance of his relations and friends, he had expelled from the kingdom; and her camp lay at a small distance from his. To him Pompey applied to be permitted to take refuge in Alexandria, and to be protected in his calamity by his powerful assistance, in consideration of the friendship and amity which had subsisted between his father and him. But Pompey's deputies having executed their commission, began to converse with less restraint with the king's troops, and to advise them to act with friendship to Pompey, and not to think meanly of his bad fortune. In Ptolemy's army were several of Pompey's soldiers, of whom Gabinius had received the command in Syria, and had brought them over to Alexandria, and at the conclusion of the war had left with Ptolemy the father of the young king.

CIV. — The king's friends, who were regents of the kingdom during the minority, being informed of these things, either induced by fear, as they afterwards declared, lest Pompey should corrupt the king's army, and seize on Alexandria and Egypt; or despising his bad fortune, as in adversity friends commonly change to enemies, in public gave a favourable answer to his deputies, and desired him to come to the king; but secretly laid a plot against him, and despatched Achillas, captain of the king's guards, a man of singular boldness, and Lucius Septimius a military tribune to assassinate him. Being kindly addressed by them, and deluded by an acquaintance with Septimius, because in the war with the pirates the latter had commanded a company under him, he embarked in a small boat with a few attendants, and was there murdered by Achillas and Septimius. In like manner, Lucius Lentulus was seized by the king's order, and put to death in prison.

CV. — When Caesar arrived in Asia, he found that Titus Ampius had attempted to remove the money from the temple of Diana at Ephesus; and for this purpose had convened all the senators in the province that he might have them to attest the sum, but was interrupted by Caesar's arrival, and had made his escape. Thus, on two occasions, Caesar saved the money of Ephesus. It was also remarked at Elis, in the temple of Minerva, upon calculating and enumerating the days, that on the very day on which Caesar had gained his battle, the image of Victory which was placed before Minerva, and faced her statue, turned about towards the portal and entrance of the temple; and the same day, at Antioch in Syria, such a shout of an army and sound of trumpets was twice heard, that the citizens ran in arms to the walls. The same thing happened at Ptolemais; a sound of drums too was heard at Pergamus, in the private and retired parts of the temple, into which none but the priests are allowed admission, and which the Greeks call Adyta (the inaccessible), and likewise at Tralles, in the temple of Victory, in which there stood a statue consecrated to Caesar; a palm-tree at that time was shown that had sprouted up from the pavement, through the joints of the stones, and shot up above the roof.

CVI. — After a few days' delay in Asia, Caesar, having heard that Pompey had been seen in Cyprus, and conjecturing that he had directed his course into Egypt, on account of his connection with that kingdom, set out for Alexandria with two legions (one of which he ordered to follow him from Thessaly, the other he called in from Achaia, from Fufius, the lieutenant-general) and with eight hundred horse, ten ships of war from Rhodes, and a few from Asia. These legions amounted but to three thousand two hundred men; the rest, disabled by wounds received in various battles, by fatigue and the length of their march, could not follow him. But Caesar, relying on the fame of his exploits; did not hesitate to set forward with a feeble force, and thought that he would be secure in any place. At Alexandria he was informed of the death of Pompey: and at his landing there, heard a cry among the soldiers whom the king had left to garrison the town, and saw a crowd gathering towards him, because the fasces were carried before him; for this the whole multitude thought an infringement of the king's dignity. Though this tumult was appeased, frequent disturbances were raised for several days successively, by crowds of the populace, and a great many of his soldiers were killed in all parts of the city.

CVII. — Having observed this, he ordered other legions to be brought to him from Asia, which he had made up out of Pompey's soldiers; for he was himself detained against his will, by the etesian winds, which are totally unfavourable to persons on a voyage from Alexandria. In the meantime, considering that the disputes of the princes belonged to the jurisdiction of the Roman people, and of him as consul, and that it was a duty more incumbent on him, as in his former consulate a league had been made with Ptolemy the late king, under sanction both of a law, and a decree of the senate, he signified that it was his pleasure, that king Ptolemy, and his sister Cleopatra, should disband their armies, and decide their disputes in his presence by justice, rather than by the sword.

CVIII. — A eunuch named Pothinus, the boy's tutor, was regent of the kingdom on account of his youthfulness. He at first began to complain amongst his friends, and to express his indignation, that the king should be summoned to plead his cause: but afterwards, having prevailed on some of those whom he had made acquainted with his views to join him, he secretly called the army away from Pelusium to Alexandria, and appointed Achillas, already spoken of, commander-in-chief of the forces. Him he encouraged and animated by promises both in his own and the king's name, and instructed him both by letters and messages how he should act. By the will of Ptolemy the father, the elder of his two sons and the more advanced in years of his two daughters were declared his heirs, and for the more effectual performance of his intention, in the same will he conjured the Roman people by all the gods, and by the league which he had entered into at Rome, to see his will executed. One of the copies of his will was conveyed to Rome by his ambassadors to be deposited in the treasury, but the public troubles preventing it, it was lodged with Pompey: another was left sealed up, and kept at Alexandria.

CIX. — Whilst these things were debated before Caesar, and he was very anxious to settle the royal disputes as a common friend and arbitrator; news was brought on a sudden that the king's army and all his cavalry were on their march to Alexandria. Caesar's forces were by no means so strong that he could trust to them, if he had occasion to hazard a battle without the town. His only resource was to keep within the town in the most convenient places, and get information of Achillas's designs. However he ordered his soldiers to repair to their arms; and advised the king to send some of his friends, who had the greatest influence, as deputies to Achillas and to signify his royal pleasure. Dioscorides and Serapion, the persons sent by him, who had both been ambassadors at Rome, and had been in great esteem with Ptolemy the father, went to Achillas. But as soon as they appeared in his presence, without hearing them, or learning the occasion of their coming, he ordered them to be seized and put to death. One of them, after receiving a wound, was taken up and carried off by his attendants as dead: the other was killed on the spot. Upon this, Caesar took care to secure the king's person, both supposing that the king's name would have great influence with his subjects, and to give the war the appearance of the scheme of a few desperate men, rather than of having been begun by the king's consent.

CX. — The forces under Achillas did not seem despicable, either for number, spirit, or military experience; for he had twenty thousand men under arms. They consisted partly of Gabinius's soldiers, who were now become habituated to the licentious mode of living at Alexandria, and had forgotten the name and discipline of the Roman people, and had married wives there, by whom the greatest part of them had children. To these was added a collection of highwaymen and free-booters, from Syria, and the province of Cilicia, and the adjacent countries. Besides several convicts and transports had been collected: for at Alexandria all our runaway slaves were sure of finding protection for their persons on the condition that they should give in their names, and enlist as soldiers: and if any of them was apprehended by his master, he was rescued by a crowd of his fellow soldiers, who being involved in the same guilt, repelled, at the hazard of their lives, every violence offered to any of their body. These by a prescriptive privilege of the Alexandrian army, used to demand the king's favourites to be put to death, pillage the properties of the rich to increase their pay, invest the king's palace, banish some from the kingdom, and recall others from exile. Besides these, there were two thousand horse, who had acquired the skill of veterans by being in several wars in Alexandria. These had restored Ptolemy the father to his kingdom, had killed Bibulus's two sons; and had been engaged in war with the Egyptians; such was their experience in military affairs.

CXI. — Full of confidence in his troops, and despising the small number of Caesar's soldiers, Achillas seized Alexandria, except that part of the town which Caesar occupied with his troops. At first he attempted to force the palace; but Caesar had disposed his cohorts through the streets, and repelled his attack. At the same time there was an action at the port: where the contest was maintained with the greatest obstinacy. For the forces were divided, and the fight maintained in several streets at once, and the enemy endeavoured to seize with a strong party the ships of war; of which fifty had been sent to Pompey's assistance, but after the battle in Thessaly had returned home. They were all of either three or five banks of oars, well equipped and appointed with every necessary for a voyage. Besides these, there were twenty-two vessels with decks, which were usually kept at Alexandria, to guard the port. If they made themselves masters of these, Caesar being deprived of his fleet, they would have the command of the port and whole sea, and could prevent him from procuring provisions and auxiliaries. Accordingly that spirit was displayed, which ought to be displayed when the one party saw that a speedy victory depended on the issue, and the other their safety. But Caesar gained the day, and set fire to all those ships, and to others which were in the docks, because he could not guard so many places with so small a force; and immediately he conveyed some troops to the Pharos by his ships.

CXIL — The Pharos is a tower on an island, of prodigious height, built with amazing works, and takes its name from the island. This island lying over against Alexandria forms a harbour; but on the upper side it is connected with the town by a narrow way eight hundred paces in length, made by piles sunk in the sea, and by a bridge. In this island some of the Egyptians have houses, and a village as large as a town; and whatever ships from any quarter, either through mistaking the channel, or by the storm, have been driven from their course upon the coast, they constantly plunder like pirates. And without the consent of those who are masters of the Pharos, no vessels can enter the harbour, on account of its narrowness. Caesar being greatly alarmed on this account, whilst the enemy were engaged in battle, landed his soldiers, seized the Pharos, and placed a garrison in it. By this means he gained this point, that he could be supplied without danger with corn and auxiliaries: for he sent to all the neighbouring countries, to demand supplies. In other parts of the town, they fought so obstinately, that they quitted the field with equal advantage, and neither were beaten (in consequence of the narrowness of the passes); and a few being killed on both sides, Caesar secured the most necessary posts, and fortified them in the night. In this quarter of the town was a wing of the king's palace, in which Caesar was lodged on his first arrival, and a theatre adjoining the house which served as for citadel, and commanded an avenue to the port and other docks. These fortifications he increased during the succeeding days, that he might have them before him as a rampart, and not be obliged to fight against his will. In the meantime Ptolemy's younger daughter, hoping the throne would become vacant, made her escape from the palace to Achillas, and assisted him in prosecuting the war. But they soon quarrelled about the command, which circumstance enlarged the presents to the soldiers, for each endeavoured by great sacrifices to secure their affection. Whilst the enemy was thus employed, Pothinus, tutor to the young king, and regent of the kingdom, who was in Caesar's part of the town, sent messengers to Achillas, and encouraged him not to desist from his enterprise, nor to despair of success; but his messengers being discovered and apprehended, he was put to death by Caesar. Such was the commencement of the Alexandrian war.

Books in the Foundations of Fantasy Roleplaying Series

CP0001
The Adventures of Gil Blas of Santillane
Alain-René Lesage

CP0002
Told by the Death's Head
Mór Jókai

CP0003
Three Northern Love Stories, and Other Tales
unknown

CP0004
The Life and Adeventures of Guzman D'Alfarache
Mateo Aleman

CP0005
Gyfford of Weare
Jeffery Farnol

CP0006
The Wood Beyond the World
William Morris

CP0007
The Seven Tablets of Creation, or the Enuma Elish
unknown

CP0008
Under the Red Robe
Stanley Weyman

CP0009
The Moon Pool
A. Merritt

CP0010
The Three Impostors, or the Transmutations
Arthur Machen

CP00011
Arthurian Romances
Chrétien de Troyes

CP0012
Song of Roland
unknown

CP0013
The Gallic and Civil Wars
Julius Caesar

CP0014
The Life and Adventures of Lazarillo de Tormes
"Don Diego Huertada de Mendoza"

CP0015
The Cualgne Cattle-Raid
unknown

*

*

*

with more
forthcoming

Made in the USA
Monee, IL
04 November 2026

67383414R00164